Lecture Notes in Biomathematics

Managing Editor: S. Levin

34

Richard Collins
Terry J. van der Werff

Mathematical Models of the Dynamics of the Human Eye

Springer-Verlag
Berlin Heidelberg New York

Lecture Notes in Biomathematics

Vol. 1: P. Waltman, Deterministic Threshold Models in the Theory of Epidemics. V, 101 pages. 1974.

Vol. 2: Mathematical Problems in Biology, Victoria Conference 1973. Edited by P. van den Driessche. VI, 280 pages. 1974.

Vol. 3: D. Ludwig, Stochastic Population Theories. VI, 108 pages. 1974.

Vol. 4: Physics and Mathematics of the Nervous System. Edited by M. Conrad, W. Güttinger, and M. Dal Cin. XI, 584 pages. 1974.

Vol. 5: Mathematical Analysis of Decision Problems in Ecology. Proceedings 1973. Edited by A. Charnes and W. R. Lynn. VIII, 421 pages. 1975.

Vol. 6: H. T. Banks, Modeling and Control in the Biomedical Sciences. V, 114 pages. 1975.

Vol. 7: M. C. Mackey, Ion Transport through Biological Membranes, An Integrated Theoretical Approach. IX, 240 pages. 1975.

Vol. 8: C. DeLisi, Antigen Antibody Interactions. IV, 142 pages. 1976.

Vol. 9: N. Dubin, A Stochastic Model for Immunological Feedback in Carcinogenesis: Analysis and Approximations. XIII, 163 pages. 1976.

Vol. 10: J. J. Tyson, The Belousov-Zhabotinskii Reaktion. IX, 128 pages. 1976.

Vol. 11: Mathematical Models in Medicine. Workshop 1976. Edited by J. Berger, W. Bühler, R. Repges, and P. Tautu. XII, 281 pages. 1976.

Vol. 12: A. V. Holden, Models of the Stochastic Activity of Neurones. VII, 368 pages. 1976.

Vol. 13: Mathematical Models in Biological Discovery. Edited by D. L. Solomon and C. Walter. VI, 240 pages. 1977.

Vol. 14: L. M. Ricciardi, Diffusion Processes and Related Topics in Biology. VI, 200 pages. 1977.

Vol. 15: Th. Nagylaki, Selection in One- and Two-Locus Systems. VIII, 208 pages. 1977.

Vol. 16: G. Sampath, S. K. Srinivasan, Stochastic Models for Spike Trains of Single Neurons. VIII, 188 pages. 1977.

Vol. 17: T. Maruyama, Stochastic Problems in Population Genetics. VIII, 245 pages. 1977.

Vol. 18: Mathematics and the Life Sciences. Proceedings 1975. Edited by D. E. Matthews. VII, 385 pages. 1977.

Vol. 19: Measuring Selection in Natural Populations. Edited by F. B. Christiansen and T. M. Fenchel. XXXI, 564 pages. 1977.

Vol. 20: J. M. Cushing, Integrodifferential Equations and Delay Models in Population Dynamics. VI, 196 pages. 1977.

Vol. 21: Theoretical Approaches to Complex Systems. Proceedings 1977. Edited by R. Heim and G. Palm. VI, 244 pages. 1978.

Vol. 22: F. M. Scudo and J. R. Ziegler, The Golden Age of Theoretical Ecology: 1923–1940. XII, 490 pages. 1978.

Vol. 23: Geometrical Probability and Biological Structures: Buffon's 200th Anniversary. Proceedings 1977. Edited by R. E. Miles and J. Serra. XII, 338 pages. 1978.

Vol. 24: F. L. Bookstein, The Measurement of Biological Shape and Shape Change. VIII, 191 pages. 1978.

Vol. 25: P. Yodzis, Competition for Space and the Structure of Ecological Communities. VI, 191 pages. 1978.

Lecture Notes in Biomathematics

Managing Editor: S. Levin

34

Richard Collins
Terry J. van der Werff

Mathematical Models of
the Dynamics
of the Human Eye

Springer-Verlag
Berlin Heidelberg New York 1980

Authors

Richard Collins
Université Paul Sabatier
Génie Biologique et Médical
Centre Hospitalier Universitaire de Toulouse-Rangueil
Toulouse, France

Terry J. van der Werff
Department of Biomedical Engineering
University of Cape Town and Groote Schuur Hospital
Cape Town, South Africa

AMS Subject Classifications (1980): 76 Z 99, 92-02, 92 A 09

ISBN 3-540-09751-1 Springer-Verlag Berlin Heidelberg New York
ISBN 0-387-09751-1 Springer-Verlag New York Heidelberg Berlin

Printing and binding: Beltz Offsetdruck, Hemsbach/Bergstr.
2141/3140-543210

PREFACE

A rich and abundant literature has developed during the last half century
dealing with mechanical aspects of the eye, mainly from clinical and experimental
points of view. For the most part, workers have attempted to shed light on the
complex set of conditions known by the general term *glaucoma*. These conditions
are characterised by an increase in intraocular pressure sufficient to cause degeneration of the optic disc and concomitant defects in the visual field, which,
if not controlled, lead to inevitable permanent blindness. In the United States
alone, an estimated 50,000 persons are blind as a result of glaucoma, which strikes
about 2% of the population over 40 years of age (Vaughan and Asbury, 1974).

An understanding of the underlying mechanisms of glaucoma is hindered by the
fact that elevated intraocular pressure, like a runny nose, is but a symptom which
may have a variety of causes. Only by turning to the initial pathology can one
hope to understand this important class of medical problems.

Invaluable laboratory information has been obtained through skillful measurements by numerous researchers on the eyes of dogs, cats, rabbits and monkeys, as
well as on enucleated human eyes. Bioengineers have aided in the development of
the delicate measurement systems for recording pressure and fluid flow within the
(deformable) eyeball. To a far lesser extent, applied mathematicians have proposed
analytic models for the response of the eye to mechanical and biological perturba-
tions. Their investigations of noninvasive techniques of pressure measurement by
tonometry and ophthalmodynamometry may lead to improvements in the calibration and
interpretation of such recordings, particularly in early glaucoma.

This monograph presents a unified view of ocular dynamics for use by researchers pursuing quantitative investigations into the mechanical functions of the
human eye. It includes some of our heretofore unpublished results. We have been
stimulated to write this monograph by the relative paucity of bioengineering
literature dealing with the eye, in stark contrast with that treating other physiological systems, such as cardiovascular, nervous, and skeletal.

Our approach is to present simple, preliminary mathematical models which
describe volume and pressure changes within the eye as functions of measurable
ocular properties. Theoretical results are compared where possible with experimental studies. At the present stage of development, these initial results serve
to corroborate *the principle of ocular equilibrium states which forms the basis
of the mathematical analysis and constitutes a framework for the development of
further refinements of the theory and experiments.*

We hope this survey of ocular dynamics will attract more applied mathematicians,
applied mechanicians, bioengineers, and biologists to contribute further to
the formulation of realistic quantitative descriptions for use in clinical testing
and diagnosis. Indeed, the eye is an imminently accessible physiological system
ripe for interdisciplinary study. A glossary of physiological terms and mathemati-

cal notations employed are supplied at the end of the monograph as an aid to readers
of differing technical specialities.

Our work on the eye began independently in the late 1960's at the RAND Corporation in Santa Monica, California. We joined forces several years later. For attracting us to this field and guiding our initial efforts, we thank Carl Gazley, Joe Gross, and Jerry Aroesty. We are indebted to Mrs. Val Sharkey who prepared the typescript and to Miss Jenny Pitt who aided with some of the figures. We are grateful to Professors Anders Bill, Albert Kobayashi, Justin van Selm, and, particularly, Shelly Weinbaum who read the manuscript and made constructive suggestions.

14 December 1979   Richard Collins
                                             Toulouse, France

                                             Terry J. van der Werff
                                             Cape Town, South Africa

TABLE OF CONTENTS

1. INTRODUCTION 1
    1. Anatomy of the eye 2
    2. Regulation of intraocular pressure 5
    3. Units and conventions 6

2. VOLUME CHANGES IN THE EYE 8
    1. Ocular circulation 8
    2. Pressure - flow relation for the vascular bed 12
    3. Pressure-volume relation for the vascular bed 14
    4. Aqueous humour dynamics 19
        A. Aqueous formation rate 21
        B. Aqueous outflow 23
    5. Ocular rigidity function 29
        A. Friedenwald's formulation 30
        B. Other empirical formulations 31
        C. Linear elastic model 32
        D. Nonlinear elastic model 34
        E. Comparison of ocular rigidity functions 36
        F. Viscoelasticity of corneo-scleral envelope 38
    6. Causal relationships (influence diagram) 40
    7. The Standard Eye 42
    8. A summary of normal values and relations 43

3. GENERAL TIME-DEPENDENT MODEL 45
    1. Governing equations 45
    2. Approximate analytical solutions 48
        A. Steady state pressure pulsations 48
        B. Non-steady state pressure pulsations 49
    3. Numerical solutions 50

4. NEURAL CONTROL OF THE INTRAOCULAR PRESSURE 54

5. MEASUREMENT TECHNIQUES 56
    1. Measurement of intraocular pressure 56
        A. Tonometry 56
        B. Tonography 57
        C. Perilimbal suction cup method 62

6. RELATION OF OCULAR DYNAMICS TO THE CEREBRAL CIRCULATION 64
    1. Clinical methods of assessing the cerebral circulation 65
        A. Ophthalmodynamometry 65
        B. Ophthalmodynamography 66
        C. Ocular pulse analysis 68
        D. Carotid compression 69
        E. Other methods 70

7. SUMMARY AND CONCLUSIONS 72

GLOSSARY 73

MATHEMATICAL NOTATION 78

REFERENCES 82

SUBJECT INDEX 95

LIST OF FIGURES

1.1 Schematic of the eye 2

1.2 Structure of the cornea 3

1.3 Vascular supply to the eye 4

1.4 Vascular bed of the optic nerve 5

2.1 Ocular circulation: the main arteries and veins of the eye 8

2.2 Forces acting on a blood vessel 15

2.3 Comparison of vascular pressure-volume relationship with identical slopes at V_{ao} = 30 µl. 19

2.4 Structure of eye near Schlemm's canal 21

2.5 Forces acting on the corneo-scleral envelope 33

2.6 Comparison of ocular rigidity functions with identical slopes at P = 15.5 mmHg 37

2.7 Comparison of ocular rigidity functions with constants as given by the respective authors 38

2.8 Influence diagram 40

3.1 Effect of variations in the aqueous outflow parameter a_2 on the intaocular pressure 51

3.2 Effect of variations in the ocular rigidity K_t on the intraocular pressure ..52

3.3 Effect of variations in the cutoff pressure P_c on the intraocular pressure ..52

4.1 Arterial vasomotor control in the iris 54

5.1 The three main types of tonometers 56,57

5.2 Sample tonogram 58

5.3 Model results for Schiøtz indentation tonography 61

5.4 Model results for suction cup tonography 63

6.1 Upper systemic arteries showing common sites of occlusion 64

6.2 Determination of systolic and diastolic pressures in Hager ophthalmo-dynamography 67

6.3 Schematic indication of collateralisation during carotid occlusion 69

CHAPTER 1. INTRODUCTION

The eye is a complex physiological system controlled by mechanical, biochemical, and neurological factors, which, under normal conditions, assure the stability
and regulation of the intraocular pressure. This stability and regulation are
essential for the maintenance of the eye's visual functions and for the nourishment
of its tissues.

The aim of this monograph is to present an analysis of the eye's overall
mechanical behaviour, with a particular view to examining the dynamics of intraocular pressure variations. The visual, neurological, and muscular functions of
the eye are not considered at this level. Nor are the complete physiology and bio-
chemistry of the eye reviewed.

The main thrust is towards a quantitative formulation of the variations of
intraocular pressure and blood vessel pressures as functions of the mechanical
properties of the eye and the external conditions to which it is subjected. The
detailed extensions of this to the full range of pathological situations is straightforward.

The equilibrium state of the eye, in which the intraocular pressure is maintained at a nearly constant level 15-20 mmHg above ambient pressure (one of the
highest of any organ in the body), is achieved through a delicate balance of fluid
flow into and out of the eye. Under pathological conditions, the outflow resistance
may increase, leading initially to accumulation of fluid in the eye, with a concomitant increase of pressure, which may result in glaucoma. The corneo-scleral
envelope stretches in response to changes in the intraocular pressure, followed by
a process of stress relaxation.

All eyes exhibit similar general dynamic characteristics, even though there is
no such thing as an average eye. Yet, in order to formulate mathematical models of
the eye, or of any other biological or physiological system for that matter, assumptions must be made, experimental averages obtained, and relationships postulated.
All of these we shall do.

A great body of literature exists concerning the various relationships in the
eye, but the results are not well correlated. Most of the research has necessarily
been conducted on animals, usually cats and rabbits, whose eyes do not have precise-
ly the same anatomical structure as human eyes. Research on humans is usually
limited to eyes before and after enucleation for pathological reasons, neither of
which circumstances can be considered "normal".

Nonetheless, the eye is susceptible to mathematical analysis. This monograph
summarises what is known or postulated about the most important physiological
processes and relationships in the eye, based on data reported in the literature.
No new experimental evidence is reported here, although original analytical work is.
Conflicting results are noted and justifications given for particular choices for
inclusion in the theoretical model. In addition, we indicate to which species the

experiments refer, e.g. human, cat, or rabbit. We subsequently formulate a general time-dependent model of the eye and apply it to several illustrative experimental and clinical situations.

This preliminary theory offers unifying concepts which may serve as a framework for a controlled programme of experimentation. The theory would then be subject to further refinements based on new experimental findings.

1. Anatomy of the eye

The eye is a nearly spherical fluid-filled elastic shell with an anterior (corneal) bulge. (Figure 1.1) The compartments of the eye comprise the corneoscleral covering, the vascular bed, the vitreous, the lens, the iris, the retina, and the anterior and posterior chambers filled with aqueous humour. The eyeball itself occupies only about 20% of the volume of the eye socket, the remaining space being filled with muscles and ligaments, blood vessels, nerves and fatty tissue.

In man, the external diameter of the eye globe is about 24 mm, giving an external volume of about 7,000 µl. The thickness of the eye's covering - the cornea in front and the sclera behind - varies between 0.3 and 1.3 mm, so that the internal volume of the eye is approximately 6,000 µl. Table I gives approximate comparative information for human, cat, and rabbit eyes.

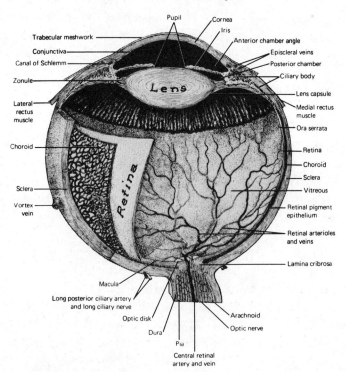

Figure 1.1. Schematic of the eye. From Vaughan and Asbury (1977) - used by permission.

TABLE I. EYE DIMENSIONS

	Human	Cat	Rabbit
Eye diameter	24 mm	21 mm	16 mm
Eye volume (external)	7,000 µl	5,000 µl	2,000 µl
Eye volume (internal)	6,000 µl	4,000 µl	1,500 µl
Anterior chamber volume	180-280 µl	300-690 µl	285 µl
Posterior chamber volume	60 µl		57 µl

The cornea - the window of the eye - is a 5-layer "sandwich" (Figures 1.2 and 2.4a) 0.6-1.0 mm thick with normal radii of curvature of 7.8 mm (anterior surface) and 7.0 mm (posterior surface). Since transparency is essential to fulfill the eye's visual function, blood (which is opaque) cannot nourish the tisssues of the eye lying within the light pathway or visual axis. The cornea and lens are nourished by a transparent fluid, aqueous humour, which is 98.1% water, contains sugar and amino acids, and is secreted by the ciliary body from the circulating blood. Aqueous humour fills the anterior chamber (2.5-4% of the total volume of the eye globe), and, as will be seen shortly, its inflow and outflow dominate the intraocular volume changes and hence the intraocular pressure.

Light enters the transparent cornea (index of refraction I.R. = 1.376), traverses the anterior chamber filled with aqueous humour (I.R. = 1.336), and is focussed by the lens (I.R. = 1.42), under the control of the ciliary muscles, to form an image upon the retina, which is linked via the optic nerve to the brain, where the image is interpreted.

The ciliary body and retina are nourished via the uveal and retinal circula-

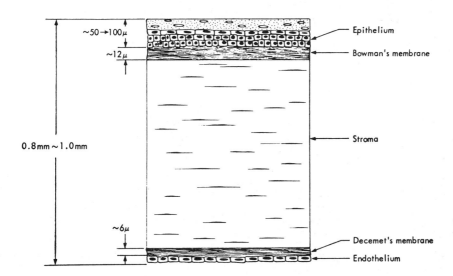

Figure 1.2. Structure of the cornea. From Mow (1968) - used by permission.

tions, which branch off the internal carotid artery.(Figure 1.3) Maintenance of
this blood supply is of prime importance, as vision is lost if the central retinal
artery which supplies the inner layers of the retina is occluded. The *metabolic
requirements of the retina* are of such high order that even slight vascular disturbances lead to marked interference with its visual function (Adler, 1965, p. 308).
Under a sudden aircraft manoeuvre, for example, high centrifugal forces may develop
which prevent blood from reaching the retina and/or the brain, resulting in pilot
'blackout'. Momentary loss of vision due to interruption of the retinal blood
supply is known to occur during spasm of the central retinal artery or partial
occlusion of the internal carotid artery. This condition of "amaurosis fugax" may
signal an impending cerebrovascular accident.

Although a sufficiently elevated intraocular pressure may eventually restrict
the retinal blood supply, ophthalmologists generally accept that glaucoma is associated with a more critical condition which intervenes at a relatively lower, but
still abnormally high, level of intraocular pressure. Thus, without necessarily
impeding the retinal blood flow, sufficiently high, prolonged intraocular pressure
may result in compression of the axonal blood vessels, particularly at the level of
the lamina cribrosa, where the optic nerve bundle passes through a "sieve-like"
opening in the sclera, on its way to the brain (Figure 1.1).

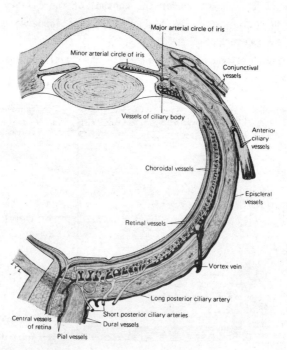

Figure 1.3. Vascular supply to the eye. From Vaughan and Asbury (1977) - used by
permission.

The effect of high, prolonged intraocular pressure may be to cause the lamina cribrosa depression to bottom out into a "bean pot" configuration. In the process, certain blood vessels, in particular the pial artery and vein which supply the periphery of the optic nerve, may collapse under the excessive external pressure. At higher levels of intraocular pressure, the blood flow may be further impaired in the central retinal artery and vein which supply the core of the nerve bundle (Figure 1.4).

At still higher levels of intraocular pressure, these vessels become susceptible to collapse at the point where they enter the lamina cribrosa depression. Anteriorly, these vessels are girdled in a relatively stiff extension of the scleral envelope against which the blood vessels will be squeezed. Posterior to this transition section, however, this stiff outer sheath gives way to the more flexible dura which deforms more readily in response to the high intraocular pressure transmitted through the lamina cribrosa. Deformation of this dural sheath tends to relieve partially the intraocular pressure and consequently to retard, if not preclude, the collapse of these important vessels. Histological studies of axonal damage appear to correlate well with the collapse of axonal flow in this localised entry region of the lamina cribrosa.

Visual loss resulting from glaucoma cannot be restored, and present therapy is directed towards maintaining the intraocular pressure at a level which does not cause further damage to the optic nerve.

2. Regulation of intraocular pressure

The intraocular pressure varies mechanically as the corneo-scleral envelope deforms in response to fluid flows through the eye, namely those of aqueous humour, blood, and vitreous. The vitreous humour does not play a role in short term intra-

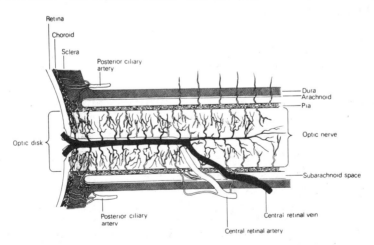

Figure 1.4. Vascular bed of the optic nerve. From Vaughan and Asbury (1977) - used by permission.

ocular pressure fluctuations. Its volume is essentially constant since its turnover time is of the order of weeks (Kronheim *et al.*, 1976), although certain low molecular weight substances within it may have a much shorter turnover time. For chronically high intraocular pressures, vitreous dynamics may become important.

Of the aqueous humour and blood flows, the former is more significant for the eye's mechanical behaviour. The two rates are, however, closely linked, and the *regulation of the intraocular pressure relates essentially to the inflow and outflow of aqueous humour and blood*. The intraocular pressure remains steady only if the net flow (outflow minus inflow) of these fluids is balanced by a corresponding variation in the volume enclosed by the corneo-scleral envelope. The factors relating the intraocular volume and pressure to the flow rates of aqueous and blood will be examined in detail in Chapter 2.

Gravity may also influence the intraocular pressure. Experiments by Anderson and Grant (1973) indicated that the intraocular pressure rises by an average of 1 to 2 mmHg as an individual shifts from a sitting to a recumbent position. Such a change is not surprising, as the blood pressure in the eye should increase as the eye approaches the level of the heart. The unexpected and still unexplained feature of this simple manoeuvre is that the intraocular pressure change is immediate and persistent upon change of position. In Chapter 2, we will see that the corneo-scleral envelope of the eye has an inherently viscoelastic character which would lead one to predict a relaxation of the pressure perturbation back towards the initial state over several minutes as a new equilibrium is struck between the aqueous secretion and outflow. The dynamic model formulated in Chapter 3 should eventually shed light upon this response, which may have clinical diagnostic applications. Anderson and Grant (1973) pointed out that, in spite of the wide scatter between different individuals, these pressure changes are accentuated in patients with glaucoma.

3. Units and conventions

In the course of this monograph many experimental values will be taken from the literature for comparative and developmental purposes. Unless otherwise explicitly stated, we shall use the following mixture of metric and traditional units:

length	- mm
volume	- mm^3 = μl
flow rate	- cm^3/min = ml/min
pressure	- mmHg
mass	- gm

Logarithms denoted by "ln" are taken to the base e(= 2.71828...), whereas logarithms denoted by "log" are taken to the base 10. Natural logarithms (ln) are preferred mathematically, but common logarithms (log) persist in ophthalmic use

because of their inclusion in Friedenwald's formula. There is a direct relationship between the two types of logarithms:

$$\ln x = M \log x, \qquad\qquad (1.1)$$

where x is the quantity whose logarithm is being taken, and M is a conversion
factor of value $M = \ln 10 = 2.302585...$

Constants will be uniformly defined as positive, so that their numerical
contribution is always explicitly indicated by their accompanying sign in an
equation.

Variables, however, may be either positive or negative, depending upon the
particular situation. For example, ΔV will always represent the volume change,
i.e. the final volume minus the initial volume. If the volume falls during the
condition being studied, ΔV is negative, whereas if it rises, ΔV is positive.
This convention is particularly advantageous in complicated equations where the
direction of change (positive or negative) of a specific variable is not immediately
obvious. When the equation is solved for the variable, the positive or negative
sign attached to its numerical value then automatically indicates the direction of
change.

In this chapter we present a thorough review of those aspects of the ophthalmic literature which concern the three principal factors affecting the volume of the
eye: vascular dynamics; aqueous humour dynamics; and the viscoelastic properties of
the corneo-scleral shell.

These factors are interdependent, so the chapter closes with a description of
their causal relationships, illustrated by an influence diagram.

1. Ocular circulation

In man, blood reaches the eye through the ophthalmic artery, a branch of the
internal carotid artery (Fig. 2.1), whereas in cats and rabbits, the principal
experimental animals, the blood supply originates in both the internal and the
external carotid arteries (Adler, 1965; Bill, 1970). Man's eye comprises two
separate circulatory systems, retinal and uveal, and both emanate from the ophthalmic artery. The retinal system, formed from the central retinal artery, supplies
the inner layers of the retina and is almost perfectly autoregulatory (Bill, 1975).
The uveal (or choroidal) system, formed from the anterior and posterior ciliary
arteries, supplies the outer layers of the retina and the outer coats of the eye.
Excellent drawings and photographs of the vasculature of the cat eye may be found in

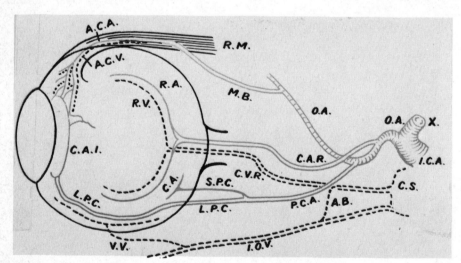

Figure 2.1.   Ocular circulation: the main arteries and veins of the eye.   AB = anastomosing branch; ACA = anterior ciliary artery; ACV = anterior ciliary
vein; CA = ciliary artery; CAI = circulus arteriosus iridis major;
CAR = central retinal artery; CS = cavernous sinus; CVR = central
retinal vein; ICA = internal carotid artery; IOV = internal ophthalmic
vein; LPC = long posterior ciliary artery; MB = muscular branch; OA =
ophthalmic artery; PCA = posterior ciliary artery; RA = retinal artery;
RM = recti muscle; RV = retinal vein; SPC = short posterior ciliary
artery; VV = vortex vein; X = point of narrowing of internal carotid
artery.   From Duke-Elder (1926) - used by permission.

Wong and Macri (1964). A comprehensive description of the arteries of the orbit in humans has been given by Hayreh (1963).

The changes in intraocular pressure brought about by variations in the volume of the blood vessels are complicated. Aside from phasic variations during respiration, which do not affect the average values, skipped heart beats, intraocular pressure variations constricting the blood vessels, and nervous mechanisms must be considered.

The arteries are relatively thick-walled, distensible vessels and tend to damp out fluctuations in the intraocular pressure - a pulse pressure of 40 mmHg in the retinal artery during cardiac systole produces a variation of only 1-2 mmHg in the intraocular pressure - whereas the thin-walled veins are sensitive to and subject to collapse under certain fluctuations of the intraocular pressure. Pulsations of the retinal vessels can be observed through an ophthalmoscope and recorded during tonography.

Three quantities are particularly important in ocular vascular dynamics: blood volume, blood flow rate, and blood pressure.

The blood volume of the rabbit eye was measured by Fish *et al.*(1969) using erythrocytes tagged with radioactive iron. They found a direct correlation between eye weight and ocular blood volume. The mean blood volume was 18.2 µl/eye with a mean eye weight of 2.68 gm, or 6.8 µl/gm, considerably lower than previous estimates of 200-500 µl (Adler, 1965, p. 317), although these earlier figures may also have included the volume of the blood vessel structures themselves.

In cats, Chao and Bettman (1957) determined the blood volume of the uveal system relative to that of the retinal system to be 37:1. Thus, the ocular pressure pulse is essentially due to pulsations of the uveal vessels (Adler, 1965, p. 313).

Levene (1957) used radioactive potassium to determine the blood flow through rabbit eyes. The mean flow rate through each eye was 0.85 ml/min, although two of the nine rabbits had unusually high flows. More recent experiments by Aronson *et al.* (1974) using SrCl-labelled microspheres gave the mean flow rate in rabbits as 1.3 ml/min. Bill (1962 a,b; 1967; 1970) reported uveal flow rates of 2.0-3.0 ml/min for rabbits and 1.2-1.5 ml/min for cats. The rabbits most probably had a marked vasodilation, since in unanaesthetised rabbits the uveal blood flow is about 1 ml/min (Bill, 1974).

Kaufmann *et al.* (1973) measured the blood flow rate through the eye using I^{131}-tagged albumen particles. The authors demonstrated a distinct dependence of blood flow through the eye on arterial P_{O_2} and P_{CO_2}. The highest values of blood flow through dog eyes (Q_{dog} = 1.86 ml/min/eye) were recorded for arterial gas pressures of P_{O_2} = 66 mmHg and P_{CO_2} = 42 mmHg, while the corresponding maximal values for cat eyes were Q_{cat} = 2.30 ml/min/eye for P_{O_2} = 110 mmHg and P_{CO_2} = 42 mmHg. These maximal single eye flow rates exceed by a factor of 6 the lowest values recorded for reduced levels of the arterial CO_2 partial pressure: (Q_{dog} = 0.3 mm/min for P_{O_2} = 100 mmHg and P_{CO_2} = 28 mmHg; and Q_{cat} = 0.4 ml/min for P_{O_2} = 140 mmHg and P_{CO_2} = 22 mmHg).

These significant vasodilative and vasoconstrictive effects of CO_2 and O_2, respectively, have been discussed by Bill (1955). Hickham and Frayser (1965) correlated changes in vessel calibre with arteriovenous O_2 differences in normal and pathological states on the basis of a mean retinal circulation time (RCT), derived from photographing the fundus after fluorescein injection. They demonstrated that the mean RCT increases by 25% with the inhalation of 100% O_2, confirming that O_2 constricts the retinal vessels, whereas sublingual glycerine decreases RCT by 40%, implying accelerated flow due to vessel dilation. On the other hand, inhalation of 7% CO_2 led to little change in the RCT.

When expressed with respect to the portion of tissue being perfused, the blood flow rate through the choroid of 50 ml/min/(gm choroid) is the highest in the body (Bill, 1967). Since most of this flow nourishes the retina, the flow calculated on the basis of the tissues supplied becomes about 10 ml/min/(gm choroid and retina). Using a Krypton-85 gas clearance technique, Wilson *et al.* (1973) estimated the choroidal flow in rabbits to be 16.6 ml/min/(gm choroid).

The blood flow rate of the retinal circulation is considerably less than that of the uveal circulation. By means of an oxygen uptake method, Anderson and Saltzmann (1964) found the retinal blood flow to be 1.72 ml/min/(gm retina) in humans, an order of magnitude lower than choroidal flow rates based on tissue perfused and two orders of magnitude lower based on total flow.

Use of 50 μ-diameter nuclide-labelled microspheres enabled Weiter *et al.* (1973) to assess the relative blood flow to different regions of the eye. They estimated that 65% of the total intraocular blood flows into the choroid, this being 40 to 70 times that to the retina.

Flow velocities in the retinal vessels have been measured by laser Doppler techniques by Tanaka, Riva and Ben-Sira (1974). In a human subject they found the blood velocities to be 19 mm/sec in a 160 μm vein, 16 mm/sec in a 130 μm vein, and 22 mm/sec in a 100 μm artery.

The consistency of the blood volume and flow rate studies can be partially verified by examining the retinal circulation time, which should be approximately equal to the total blood volume divided by the volume flow rate. For humans this has been variously given as 1.3 sec by Suvanto, Reissel, and Himanka (1960) and 4.7 by Hickham and Frayser (1965). The theoretical figure for the rabbit eye would be 6.8/5.3 = 1.3 sec, based on the data given earlier in this section.

The arterial pressure in the eye is somewhat lower than in other major arteries, such as the brachial artery. In 1922 Magitot and Bailliart estimated that the central retinal artery pressure in the normal human eye is 70 mmHg during systole and 35 in diastole. More recent results by Cole (1966) gave similar values of 75/35 for the uvea. It is difficult to measure these pressures accurately by noninvasive techniques, and it has been pointed out that the systolic pressure as determined by ophthalmodynamometry is 14 to 17 mmHg higher than the normal central retinal artery systolic pressure (Bill, 1963a; van der Werff, 1972).

Perry and Rose (1958) found the average absolute difference between retinal and brachial diastolic pressures to be 20.4 mmHg in 112 adults, two-thirds of whom had some degree of systemic arterial hypertension or neurological vascular disease. This compares favourably with the 17.4 mmHg difference between the ophthalmic and brachial arteries found by Borrás, Méndez and Martínez (1969) in 15 test cases. This difference is due almost entirely to the higher systolic pressure in the brachial artery, as the brachial diastolic pressure is only 5 mmHg higher.

Magitot (1922) promulgated two physiological principles governing the blood pressure in the eye: (1) the arterial pressure always exceeds the intraocular pressure; but (2) the venous pressure is not always greater than or equal to the intraocular pressure. There also seems to be no direct correlation between changes in the mean arterial pressure and the mean intraocular pressure (Drance, 1961). There does exist, however, a sensitive relationship between the intraocular pressure and the venous pressures.

Macri (1964) used microcannula techniques to investigate the anterior ciliary venous, circle of Hovius, and choroidal venous pressures as functions of the intraocular pressure in cats and reported highly significant correlations between the venous and intraocular pressures. One must be extremely careful, however, not to use his regression equations incorrectly in a mathematical model. Thus, when he elevated the intraocular pressure by acute infusion of 0.9% NaCl into the anterior chamber, the anterior ciliary venous pressure remained constant or fell, while the circle of Hovius pressure always fell. Use of his regression equations would "predict" a rise. The explanation of this seeming paradox, of course, is that the regression equation fits one data point from each of many eyes, whereas a mathematical model requires an equation to fit many data points for *one* eye. In the case of the choroidal venous pressure, this latter experiment was performed in six eyes, and the venous pressure rose by 84% of the intraocular pressure. In a different experimental series, Macri found the average intraocular pressure to be 0.2 mmHg lower than the anterior ciliary venous pressure, 5.9 mmHg higher than the circle of Hovius venous pressure, and 2.7 mmHg higher than the choroidal venous pressure. It should be noted, however, that the average intraocular pressure in the anterior ciliary venous experiments was twice that of the other sets of experiments and the 0.2 mmHg value should be used with great caution, if at all.

The circle of Hovius pressure (P_{cH}) is important in the cat because aqueous outflow cannot occur if the intraocular pressure $P < P_{cH}$. The circle of Hovius plays the same role in cats' eyes as the episcleral venous pressure plays in human or rabbit eyes, that is, a regulatory role in the outflow of aqueous humour.

The intraocular venous pressure in man must be higher than the intraocular pressure; otherwise, the veins would collapse. Duke-Elder (1926) estimated the intraocular venous pressure to be 2 mmHg greater than the normal intraocular pressure. This compares favourably with the 2.7 mmHg difference found in cats by Macri (1964). Duke-Elder also estimated the intraocular capillary pressure to be 50 mmHg.

Experiments on albino rabbit eyes (Masket et $al.$, 1973) demonstrated a progressive fall in the transmural pressure as one proceeds from the ophthalmic artery through the choroid to the anterior uveal circulation. By continuously increasing the pressure external to the retinal and choroidal arteries (ophthalmodynamometry) and detecting flow blockage by fluorescein angiography, they estimated the average choroidal "wedge" pressure to be 14 mmHg lower than the systolic pressure in the ophthalmic artery. Direct cannulation experiments permitted comparisons between the choroidal and anterior uveal circulation and indicated a further drop of 5-20 mmHg between the two. For still higher external (intraocular) pressures, choroidal fluorescence did not extend as far as the iris vasculature, indicating continued collapse of the latter. They concluded that the "full force" generated by blood flow in the ophthalmic artery is not available to drive the intraocular uveal blood flow and conjectured that this might be due to the passage of the ciliary blood vessels through the sclera, thereby creating sites of high resistance.

Some years earlier, Best et $al.$ (1969) demonstrated that the blood flow stops when the ophthalmic artery pressure is reduced to an average of 6 mmHg above the intraocular pressure in enucleated cat eyes, terming this the "critical closing pressure".

The generally accepted equation for the interdependence of intraocular and episcleral venous pressures in man was given by Weigelin and Lobstein (1963):

$$P_v = 0.48 \ P + 3.1, \tag{2.1}$$

where P_v is the episcleral venous pressure and P is the intraocular pressure. Schimek (1964) quoted Linnér's (1959) work in which he found a mean episcleral venous pressure in man of 11.7 mmHg. This compares with a predicted value from Equation (2.1) of 10.5 at the mean normal intraocular pressure P_o of 15.1 mmHg. (Leydhecker, Akiyama and Neumann, 1958). In 14 rabbits, Kornbluth and Linnér (1955) found the mean episcleral pressure to be 8.9 mmHg. It is important to remember that Equation (2.1) is a statistical correlation between the pressures in many eyes and not a correlation determined within a single eye. Application to a particular eye may require modification of the constants appearing in Equation (2.1).

2. Pressure - flow relation for the vascular bed

The blood supply to the eye, as to all body tissues, is clearly of fundamental importance. The ocular blood supply depends upon two primary factors: the perfusion pressure; and the ocular vascular resistance. The latter is a function not only of the vascular architecture but also of the transmural pressure, i.e. the blood pressure minus the intraocular pressure, which determines the local cross-sectional area of the blood vessels. This will be the topic of the next section.

The present section summarises our knowledge of the first factor: the dependence of the ocular blood flow on the perfusion pressure, which Bill (1970) correctly defines as the difference between the arterial and venous pressures. In the eye, the vascular pressure-flow relation is complicated somewhat by the high "tissue" pressure (the intraocular pressure) and by the different blood pressures for veins inside or outside the sclera. However, the venous pressure at the site at which the vein exits from the eye (and removes itself from the direct influence of the intraocular pressure), is normally close to the intraocular pressure (cf. Equation (2.1)). Thus, it is standard practice to approximate the perfusion pressure as the difference between the arterial and intraocular pressures. This approximation will be acceptable at normal arterial and intraocular pressures but introduces considerable error if the arterial and intraocular pressures approach one another.

Bill (1962a) performed experiments on cats and rabbits to determine the relationship between the perfusion pressure and uveal flow. He concluded that an increase in intraocular pressure reduces the blood flow through the uvea and that there is an approximately linear relationship between the blood flow and the perfusion pressure, which he defined as the *femoral* arterial pressure minus the intraocular pressure! The mean arterial pressure was maintained by a compensating intravenous infusion of blood. The reader must infer the normal flows and perfusion pressures from Bill's graphs. These indicate that normal uveal blood flow is about 3.0 ml/min in rabbits and 1.2 ml/min in cats . Both Levene (1957) and Aronson *et al.* (1974) give lower flow rates for rabbits: 0.85 ml/min and 1.30 ml/min, respectively, which are consistant with Bill's (1974) value of 1 ml/min in unanaesthetised rabbits.

If the femoral pressure remained constant throughout the experiment, then the approximately linear relationship between the perfusion pressure (varied by changing the intraocular pressure) and the uveal blood flow rate could be deduced from his graphs. For cat eyes this gives:

$$Q_u = 0.012 \ (P_{fem} - P), \tag{2.2}$$

where Q_u is the uveal blood flow rate in ml/min, P_{fem} the femoral artery pressure, and P the intraocular pressure, the pressures being expressed in mmHg. For rabbit eyes, the relation is

$$Q_u = 0.015 \ (P_{fem} - P) - 0.3. \tag{2.3}$$

Both of these formulas are highly tentative, especially Equation (2.2) for which the data present a slightly sigmoidal shape. The factors multiplying the perfusion pressures in the above two equations are subject to about 50% variation for different eyes. Note too that flow for rabbit eyes, according to Equation (2.3), ceases while a positive perfusion pressure is still present. This would be analogous to the critical closing pressure as defined by Best *et al.* (1969).

This reduction of uveal blood flow at increased intraocular pressures triggers one of the feedback mechanisms in the eye. The lower blood flow reduces the secretion of aqueous humour which tends, in turn, to reduce the intraocular pressure.

Best *et al.* (1973) also produced approximately linear plots between uveal flow and perfusion pressure in cats, their perfusion pressure being defined as the difference between ophthalmic and intraocular pressures. Their results for whole blood give approximately:

$$Q_u = 0.0011 (P_{oph} - P), \tag{2.4}$$

which is an order of magnitude lower than Bill's results (Equation (2.2)). No explanation was given for such a large discrepancy.

Best and co-workers also demonstrated a significant amount of autoregulatory behaviour in ocular blood flow when the ciliary artery pressure rises above a critical value of about 60-90 mmHg, depending upon the individual eye. At this critical pressure there is a sharp transition from nonautoregulatory to autoregulatory behaviour. In effect, there is a linear relationship between flow and perfusion pressure both above and below the critical pressure, but the slopes of the relationship differ between the two regimes.

Bill (1975) noted that in healthy experimental animals retinal blood flow is affected little by changes in the perfusion pressure, i.e. autoregulation of retinal blood flow is almost perfect. On the other hand, in rabbits there seems to be little autoregulation of the blood flow in the anterior uvea as well as in the choroid. In cats and monkeys, choroidal blood flow seems to be essentially without autoregulation, but blood flow in the anterior uvea is autoregulated to a variable extent. Bill notes that total oxygen extraction from the choroid is little changed until very low flow rates are reached, in agreement with Best's observations above.

Weiter *et al.* (1973 a,b) showed an inverse nonlinear relationship between intraocular blood flow and intraocular pressure in the cat and alluded to the implications for the treatment of vascular occlusive disease and the pathogenesis of glaucoma.

3. Pressure-volume relation for the vascular bed

In order to consider the effects of blood flow changes on the intraocular pressure, a relation is needed between the arterial pressure, intraocular pressure, and the arterial volume. Ytteborg (1960a) was one of the first to draw attention to the role of the intraocular blood volume in measurements of ocular rigidity. He found that the rigidity coefficient of enucleated eyes was distinctly higher than in the same eyes *in vivo*, attributing the difference to the blood expelled from the eye during *in vivo* tonometry.

Consider a portion of blood vessel as shown in Figure 2.2. Equating the forces acting on a diameter of the vessel, the thin-wall approximation ($h \ll r$) gives

$$\sigma_a = \frac{r}{h}(P_a - P),$$ (2.5)

where σ_a is the stress in the circumferential direction, P_a the arterial pressure, P the intraocular pressure, and r and h the inner radius and thickness of the vessel, respectively. The strain is defined as

$$\varepsilon_a = \frac{r - r_o}{r_o},$$ (2.6)

where r_o is the radius at zero transmural pressure, i.e. for $P = P_a$. The stress and strain are related by the usual formula for an elastic, isotropic cylindrical tube:

$$\sigma_a = \frac{E_a}{1 - \nu_a^2}\varepsilon_a,$$ (2.7)

where E_a is the elastic modulus of the arterial wall, and ν_a Poisson's ratio. Substituting for σ_a and ε_a, we obtain

$$P_a - P = \frac{E_a h(r - r_o)}{(1 - \nu_a^2)rr_o}.$$ (2.8)

Alternatively, strain can be expressed in terms of the area:

$$\varepsilon_a = \frac{1}{2}\frac{A - A_o}{A_o} = \frac{1}{2}\frac{\Delta A}{A_o},$$ (2.9)

where A is the cross-sectional area corresponding to the radius r. This relation,

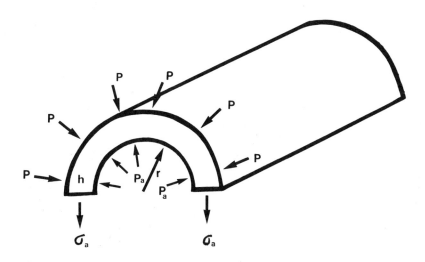

Figure 2.2. Forces acting on a blood vessel.

when substituted into Equation (2.7), gives

$$P_a - P = \frac{E_a h \Delta A}{2(1-\nu_a^2)rA_o} . \tag{2.10}$$

The incompressibility of the vessel wall yields an approximate relation for the wall thickness in thin-walled tubes:

$$h = h_o \frac{r_o}{r} . \tag{2.11}$$

Substitution of h from Equation (2.11) into (2.10) leads to

$$P_a - P = \frac{E_a h_o r_o \Delta A}{2(1-\nu_a^2)r^2 A_o} = \frac{E_a h_o \Delta A}{2(1-\nu_a^2)r_o A} . \tag{2.12}$$

If we multiply both numerator and denominator in Equation (2.12) by an effective length, we transform it into an equation relating the transmural pressure to volume changes:

$$P_a - P = K_a \frac{\Delta V_a}{V_a} , \tag{2.13}$$

with the proportionality factor

$$K_a \equiv \frac{E_a h_o}{2(1-\nu_a^2)r_o} , \tag{2.14}$$

where V_a is the blood volume at the transmural pressure (P_a-P) and ΔV_a the volume increment (V_a-V_{ao}), i.e. the difference in blood volumes between finite and zero transmural pressures.

It may be more appropriate to use as the reference condition not zero transmural pressure, but the critical closure pressure which Best *et al.* (1969, 1970) determined to be $P_a = P + \Delta P_{cc}$, where ΔP_{cc} is about 6-10 mmHg, the latter value seeming to be preferred. None of the above equations need be changed, but all variables with an "o" subscript would then refer to their values at $P_a = P + \Delta P_{cc}$. That $\Delta P_{cc} \neq 0$ further indicates the simple linear analysis above has its limitations.

Best and Blumenthal (1972) have shown that the vasomotor tone is increased by adding levaterenol bitartrate to the perfusate. One can in this way attain critical closure pressures up to 28 mmHg. They note, however, that arterial wall distensibility is apparently unaffected by vasomotor tone, in marked contrast to the smaller vessels of the systemic circulation in which distensibility decreases as vasomotor tone rises.

This linear analysis is strictly true for thin-walled tubes with constant elastic properties undergoing small strains, but these conditions are not likely to be fulfilled in the intraocular blood vessels. In addition, the intraocular blood vessels are of different types (artery, capillary, vein), of different sizes, of different compositions and elastic properties, and subjected to different transmural pressures, because of the drop in the pressure along the vascular bed.

One may relax successively two of the restrictions of this linear model; namely those of small strain and constant elastic modulus. For large strains, of the type encountered in the eye, the integrated form of Equation (2.13) should be used:

$$\Delta P_a - \Delta P \equiv (P_{a_2} - P_{a_1}) - (P_2 - P_1) = K_a \ln(V_{a_2}/V_{a_1}), \qquad (2.15)$$

where the subscripts "$_1$" and "$_2$" denote the initial and final states, respectively. One can again see the limitations of a linear analysis, since Equation (2.15) predicts smaller pressure rises for larger initial volumes, rather than the larger pressure rises as always found experimentally (cf. Best *et al.* (1971)).

This discrepancy now requires that the second restriction be lifted by recognising that the elastic modulus is not constant but depends upon the state of stress (Bergel, 1961). In fact, if one fits a curve to Bergel's data for arteries, one finds that

$$E_a = k (P_a - P)^\alpha, \qquad (2.16)$$

with $k \approx 2.5$, $\alpha \approx 1.6$, and the units of E_a being mmHg. The principal reason for the dependence of E_a on the transmural pressure is that the arterial wall is subjected to strain, the relatively stiff collagen fibres "uncoil" and pick up more of the load from the elastin fibres which carry most of the low strain load. If we now put Equation (2.16) into (2.12), we find for *small* strains

$$P_a - P = K_\alpha (P_a - P)^\alpha \frac{\Delta V_a}{V_a}, \qquad (2.17)$$

where K_α is identical to K_a except that k replaces E_a in the latter. The analogous integrated form of Equation (2.17) applicable to *large* strains is

$$\frac{1}{-\alpha+1} \left[(P_{a_2} - P_2)^{-\alpha+1} - (P_{a_1} - P_1)^{-\alpha+1} \right] = K_\alpha \ln (V_{a_2}/V_{a_1}). \qquad (2.18)$$

In contrast to Equation (2.15), this formula predicts greater pressure rises for larger initial volumes, i.e. the vascular bed becomes "stiffer" as the blood volume increases.

For highly deformable tissues, another frequently used elastic formulation is (Flügge, 1962):

$$E_{inc} = \frac{\text{incremental stress}}{\text{incremental strain}} = \frac{d\left[\frac{(P-P_a)r}{h}\right]}{dr/r}, \qquad (2.19)$$

where E_{inc} is a function of the state of strain, i.e. $E_{inc} = E_{inc}(r)$. For arterial tissue a common approximation is

$$E_{inc} = E_0 (\frac{r}{r_0})^\beta, \qquad (2.20)$$

where β is found experimentally to be about 4.5. Quantitatively similar results have been found for pure elongation of arterial and other soft biological tissue (Fung, 1967; Collins and Hu, 1972).

Integration of Equation (2.19) with (2.20) inserted gives, for thin-walled tubes,

$$P_a - P = \left(\frac{r_o}{r}\right)^2 \left[\frac{E_o h_o}{\beta r_o} \{(\frac{r}{r_o})^\beta - 1\} + (P_{ao} - P_o)\right] . \qquad (2.21)$$

The factor $E_o h_o / r_o$ can be identified as $2\rho c_o^2$, where ρ is the blood density and c_o the Moens-Korteweg pulse wave speed, i.e. the ideal speed at which a pressure pulse travels along a thin-walled artery containing an inviscid fluid. c_o is of the order of 4-10 m/s (McDonald, 1974). By introduction into Equation (2.21) of an effective length of the vascular bed, we get the pressure-volume relationship:

$$P_a - P = \left(\frac{V_{ao}}{V_a}\right) \left[\frac{2\rho c_o^2}{\beta} \left\{\frac{V_a}{V_{ao}}^{\beta/2} - 1\right\} + (P_{ao} - P_a)\right] . \qquad (2.22)$$

Best and coworkers in 1971 found that the relationship between pressure and volume changes in the intraocular vascular bed can be expressed approximately by:

$$VR \equiv \frac{\Delta(P-P_a)}{\Delta V_a} = \frac{\log \{(P_a-P)/(P_{ao}-P_o)\}}{V_a - V_{ao}} , \qquad (2.23)$$

where VR is defined as the *coefficient of vascular rigidity*. Best *et al.* found that VR is approximately 0.021 μl^{-1} for cat eyes and 0.024 μl^{-1} for human eyes.

Figure 2.3 compares the five vascular pressure-volume relationships - Equations (2.13), (2.15), (2.18), (2.22), and (2.23) - for the case where the slopes are all equal to Best's experimental value with the additional assumptions that $P_{ao} - P_o$ = 40 mmHg at V_{ao} = 30 μl, ρ = 1.055 gm/cm^3, α = 1.6, and β = 4.5. Equation (2.23) requires c_o = 367 cm/sec for consistency, and this value is in the range of arterial pulse wave speeds observed in humans, although at the low end of the range.

Several observations can be made from Figure 2.3, the most obvious being that (2.18) and (2.23) differ markedly from the other three. Since (2.23) is based on experimental results, it is clear that (2.13), (2.15), and (2.22) are not suitable, even though they each realistically give a finite value of V_a for $P_a - P = 0$, whereas (2.18) gives $V_a = 0$ (which is unrealistic) and (2.23) gives $P_a - P > 0$ for $V_a = 0$ (which is wrong). It is interesting to note that except for the "tails" the very complicated (2.22) is essentially identical to the simple linear relationship (2.13). One also notes that (2.15) gets "softer" as V_a rises, contrary to physiological observation.

Since (2.23) is a purely empirical formulation, whereas (2.18) is mathematically rigorous once the empirical form of the elasticity is assumed, and since minor adjustment in K_α can bring (2.18) more closely in line with (2.23) over the range plotted in Figure 2.3, we shall incorporate Equation (2.18) in our subsequent mathematical model.

4. Aqueous humour dynamics

The aqueous humour is a clear liquid (98.1% water) filling the anterior and posterior chambers of the eye and serves several functions. It provides the metabolic oxygen demands of the lens and a portion of the cornea, both of which are without blood vessels (Adler, 1965). By balancing formation and drainage rates, it plays the key role in maintaining the intraocular pressure. Furthermore, the aqueous humour in the anterior chamber is a component of the optical system of the eye (Moses, 1970)

The aqueous humour is slightly more dense and has a viscosity 2.5-4.0% larger than water (Moses, p. 303). Thus, it is considerably different from plasma which has a similar density but a viscosity about 50% larger than water.

The anterior chamber in man has a volume of approximately 250 μl whereas the posterior chamber volume is about 60 μl. The corresponding volumes for the common experimental animals are similar. The aqueous flow rate is about 2 μl/min so that

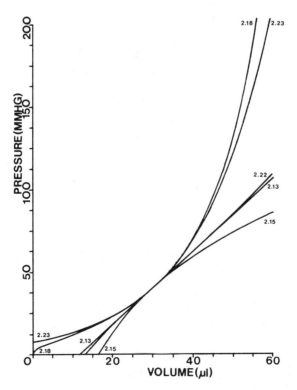

Figure 2.3. Comparison of vascular pressure-volume relationship with identical slopes at V_{ao} = 30 μl.

approximately 1% of the total aqueous humour volume is renewed each minute (Linnér, 1969; Moses, 1970; Kinsey & Bárány, 1949; Saeteren, 1960b).

The aqueous humour is important clinically because of its pressure regulatory function. If the flow of aqueous humour out of the eye is impeded, a clinical condition known as glaucoma will arise, characterised by an increase in the intra-ocular pressure. If compensatory mechanisms are not brought into play, the pressure rise can be great enough to distort the optical system of the eye, causing a degra-dation of vision and in severe glaucoma possibly leading to blindness. In the more chronic varieties of glaucoma, loss of vision is usually due to interference with axonal flow and only later with retinal circulatory shutdown.

Comprehensive reviews of the physiological and biochemical processes involved in the production of aqueous humour have been given by Kinsey (1950), Langham (1958), Cole (1966), and Bill (1975).

Kinsey and Bárány (1949) determined the rate of outflow of aqueous humour in rabbit eyes by observing the rates of disappearance of three test substances from the anterior chamber. They found that 1.07-1.30% of the total volume of aqueous humour in the anterior chamber flows out every minute and considered 1.1% to be the best estimate from the available evidence. For an anterior chamber with a volume of 250 µl, this corresponds to a flow of 2.75 µl/min. Kornbluth and Linnér (1955) quote values of 4.37 and 3.82 µl/min. Other estimates for rabbit eyes range from 1.1% to 1.9% volume turnover per minute (Adler, 1965, p. 135). For monkey eyes, Bill (1969) gives 1.61 µl/min.

After comparing their results with work on human eyes, Kinsey and Bárány (1949) suggested that the dynamics of aqueous humour of human eyes closely resemble those of the rabbit. Adler (1965, p. 135) quoted Goldmann's estimate that 1.9% of the total anterior chamber volume in man flows out every minute. Grant (1951) gives 2.4 µl/min for the normal aqueous flow in man. Schimek (1964) quotes a slightly lower rate of 2.2 µl/min, which may be more accurate since it appears that Grant overestimated the perfusion pressure.

Kinsey and his coworkers (1950) also found that the *total* normal rate of trans-port of water into and out of the anterior chamber was approximately 20% per minute of the volume of aqueous, i.e. 50 µl/min for an average rabbit eye. With 2.75 µl/min accounted for by flow, 47.25 µl/min must enter and leave the anterior chamber by internal transport. Adler (1965, p. 112) reports the percentage as 5.2% to 15.6% instead of Kinsey's 20%. The fact remains, however, that internal transport rates are much larger than the flow rates for aqueous humour.

Under steady state conditions, the intraocular pressure remains constant because the rate of aqueous production equals the rate of outflow. When the pres-sure is disturbed from the steady state, a homeostatic influence tends to return it to normal, by lowering the aqueous production rate and raising the outflow rate. These homeostatic mechanisms will be considered in the next two sections.

A. Aqueous formation rate

Aqueous humour is produced continuously from all the blood vessels within the eye, but the density of these vessels is greatest at the posterior surface of the iris and in the ciliary body (Green and Pederson, 1973). The formation process is not well understood, but it is thought to involve transport across the capillary walls.

Four mechanisms have been proposed for the formation of aqueous humour (whose composition differs considerably in the anterior and posterior chambers of the eye): ultrafiltration; dialysis; secretion; and secretion-diffusion (Adler, 1965). The osmotic pressure is not high enough for dialysis to account for the observed rates of inflow, nor can diffusion contribute significantly to aqueous formation, although both can control the chemical composition concentration gradients. Thus, some investigators support the hypotheses that aqueous humour is formed both by secretion within the cells of the ciliary epithelium (Figure 2.4a) in a manner similar to that of the secretory cells of the kidney and by ultrafiltration (Green and Pederson, 1973). Bill (1975) discusses possible shortcomings of these hypotheses.

Secretion is an active process by which water and other substances are transported across a membrane at the expense of cellular energy. The most important factor controlling the secretion of aqueous humour appears to be the intraocular pressure itself (Cole, 1966), acting in a purely mechanical way to favour or inhibit the inflow of aqueous humour. Langham (1959a) found that the formation rate of aqueous in rabbits falls as the intraocular pressure is raised, ceasing entirely when the pressure equals 90 mmHg, and that the secretion rate falls monotonically with increasing intraocular pressure. This suggests the following linear relationship between the production of aqueous humour (S_p) and the intraocular pressure:

$$S_p = C_p(P_c - P), \tag{2.24}$$

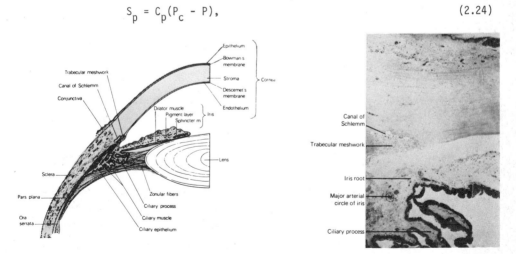

Figure 2.4. Structure of eye near Schlemm's canal. From Vaughan and Asbury (1977) - used by permission.

where P_c is the cutoff pressure and C_p is a constant. The cutoff pressure P_c is the filtration pressure required to counteract the secretory component of the aqueous production. Langham gives P_c as 90 mmHg, although P_c and the critical closing pressure of Best *et al*. (1969) are probably related. The constant C_p can be estimated under normal conditions. If S_p = 2.0 μl/min and P_o = 15.5 mmHg in the steady state for humans, then C_p = 0.027 μl/min/mmHg. Weinbaum and Goldgraben (1972) give plots of Equation (2.24) for rabbit eyes.

C_p has the same dimensions as the well-known facility of outflow C_f which we encounter in the next section. Thus, C_p might appropriately be called the "facility of aqueous production".

Equation (2.24) can be derived more rigorously on the assumption that aqueous production is directly proportional to the blood flow through the ciliary processes (Figure 2.4b). Bill (1962a) found a linear relation between the blood flow rate through the uvea and the intraocular pressure. When the intraocular pressure approaches the ophthalmic artery pressure, blood flow through the eye will cease (Best *et al*. 1969; Masket *et al*. 1973).

In experiments on living and enucleated rabbit eyes, Langham (1959a) augmented the normal rate of aqueous by infusions of fluid directly into the anterior chamber. For enucleated eyes, the pressure-infusion rate relation was linear, whereas for living eyes, the relation showed positive curvature. We fitted Langham's results for 26 living eyes of rabbits anaesthetised with urethane and found the following relation:

$$P = 21.3 + 0.5\ I + 0.9\ I^2, \qquad\qquad (2.25)$$

where I is the infusion rate in μl/min. This curve fits his data to within 2% throughout the experimental range of $21.3 \leq P \leq 56.3$ mmHg. Of course, the pressure is also influenced by changes in outflow resistance and aqueous formation at the higher infusion rates.

Green and Pederson (1973) used an *in vitro* preparation of ciliary body and iris to re-examine the relative roles of secretion (active transport) and filtration in the formation of aqueous humour at the posterior surface of the iris. They appear to be in disagreement with the interpretation of the data of Cole (1966) from which it was estimated that secretion accounts for 70% of total aqueous formation. Green and Pederson's measurements of the active Na^+ influx (indicative of secretion) *in vivo* and *in vitro* indicated that secretion accounts for about 40% of aqueous production, with the remaining, predominant contribution of about 60% deriving from ultrafiltration, a level much higher than thought earlier. Weinbaum *et al*. (1972) had earlier reached a similar conclusion by re-examining Cole's 1961 and 1962 *in vivo* and *in vitro* data. They concluded that only about one-third of the aqueous production is due to secretion and two-thirds to filtration. Cole (1966) clouded the issue by drawing the wrong quantitative conclusions from his own data, as the secretory pumps *in vivo* are nearly twice as strong as *in vitro* (Weinbaum, personal communication).

In ultrafiltration (dialysis or transport by diffusion from high to low concentration, assisted by a favourable pressure gradient), the flow rate is proportional
to the pressure difference, the membrane permeability, and its surface area. Nonetheless, as a consequence of the ultrastructure of the epithelium, the highly
involuted intercellular clefts will distend under intercellular pressure and associated fluid flow, but will collapse if there is no flow. However, the intercellular
clefts will not necessarily collapse for retrograde flow as there is still a driving
pressure to push the lateral membranes apart.

Controversy continues over whether aqueous osmotic pressure is greater or less
than that of blood, the two components being separated by the anterior surface of
the ciliary body. Green and Pederson (1973) always observed aqueous flowing *towards*
the blood. Tonicity differences can be explained by the movement of plasma proteins
at the ciliary epithelial barrier.

The osmotic pressure difference drawing aqueous into the capillaries appears to
be about 21 mmHg (Green and Pederson, 1973). Consequently, the blood pressure
necessary to overcome this and begin to produce a flow into the eye is 21 mmHg. An
additional pressure difference across the ciliary epithelium of 12-15 mmHg would be
required to obtain normal flow rates of aqueous humour of about 1.5-2 µl/min.
Finally, the intraocular pressure of 15-20 mmHg must be overcome. Summing these
three components, Green and Pederson (1973) conclude that the total capillary
pressure must be 49-57 mmHg, values falling within the range estimated by Duke-Elder
(1968).

Caution must be exercised in interpreting results on rabbit eyes traumatised
by insertion of needles which may produce a leaky blood-aqueous barrier and, hence,
a spurious non-physiological increase in ultrafiltration. Artifact-free experiments
are necessarily extremely delicate, and much work remains to be done.

B. Aqueous outflow

Two pathways have been identified for the outflow of aqueous humour (Pederson
and Green, 1973): (a) "percolation" of aqueous through a trabecular drainage network into the canal of Schlemm (Figure 2.4) to rejoin the bloodstream at the episcleral veins; and (b) a small pressure-independent uveoscleral bulk flow (Bill,
1965). The aqueous outflow process is essentially mechanical and better understood
than aqueous formation.

Aqueous humour flows from the posterior into the anterior chamber. The main
outlet from the anterior chamber is found at the filtration angle and called
Schlemm's canal (Figure 2.4). It is an annular vessel, 0.25 mm in diameter,
running around the limbus (the junction of the cornea and sclera) and separated from
the filtration angle by the trabecular network. This latter system of tiny vessels
has a resistance equivalent to that of a single tube of diameter 10-12 µm passing

through the wall of the eye (McEwen, 1958; Ruskell 1961).

From the canal of Schlemm the aqueous humour enters the episcleral veins and the intrascleral venous plexus where it mixes with blood arriving from the anterior venous plexus of the ciliary body. From this point onward, the aqueous outflow resistance is influenced by the blood flow as both aqueous and blood compete for space in the outer drainage network.

Most of the resistance to aqueous outflow is encountered in the trabecular network, but its origin is unknown. Grant (1958) added the enzyme hyaluronidase (present in synovial joints) to the anterior chamber of the enucleated eye. The outflow resistance was mysteriously reduced, visibly altering the meshwork structure. Davson (1962) suggested that the flow resistance in the trabecular network derives from mucopolysaccharide gel (hyaluronic acid) which fills perforations in the drainage network. Hyaluronidase is known to depolymerise the long-chain mucopolysaccharide molecules (Collins, 1978) which might also explain the decrease in outflow resistance observed upon its introduction into the aqueous outflow channels of the eye. As the concentration of polymerised hyaluronate in aqueous humour is not sufficient to account for this phenomenon, one suspects it also exists in the tissue of the filtration angle. Open-angle glaucoma may well be due to this hyaluronidase-sensitive barrier (Adler, 1965).

Substances also escape from the anterior chamber by diffusion into the blood vessels of the iris (Adler, 1965). Indeed, most of the total exchange of water in the anterior chamber occurs by diffusion and filtration, with only about 5% occurring by flow. Yet, the flow processes are the prime determinants in the pressure regulation within the eye.

One may liken the flow of aqueous humour through the trabecular matrix to fluid flow through a porous medium, such as soil, which is described by Darcy's equation (1856);

$$\vec{q} = \frac{-k}{\mu}\text{grad } p_m, \tag{2.26}$$

where $\vec{q}$ is the volume flow rate of fluid per unit area, p_m the pressure in the matrix, μ the fluid viscosity, and k the permeability of the matrix. The fluid and cellular structure may be considered incompressible:

$$\text{div } \vec{q} = 0. \tag{2.27}$$

Weinbaum (1965) combined these two equations into a Neumann boundary value problem. The application of well-known theorems for existence and uniqueness lead to the conclusion that, provided neither the tissue structure of the trabecular matrix nor its boundaries deform due to pressure variations in the anterior chamber and Schlemm's canal, the aqueous outflow rate is linearly proportional to the net pressure drop between the anterior chamber and the episcleral venous plexus, that is

$$S_o = C_f(P - P_v) \qquad\qquad (2.28)$$

where S_o is the total outflow of aqueous through Schlemm's canal, P the intraocular
pressure in the anterior chamber, P_v the episcleral venous pressure, and C_f an
ocular constant known as the *outflow facility*. Friedenwald was the first to state
the equation in this form, in the discussion following Grant's paper (1951). The
same relationship obtains trivially if one applies Poiseuille's or Darcy's formulas
to the flow path globally!

The outflow resistance R_f is the reciprocal of the outflow facility C_f. The
principal resistance to outflow lies in the trabecular meshwork, near the proximal
walls of Schlemm's canal (Rohen, 1960; McEwen, 1958). Poiseuille's law is

$$Q = \frac{\pi r^4}{8\mu\ell} \Delta P = \Delta P/R_f, \qquad\qquad (2.29)$$

where Q is the volume flow rate, r the radius of the vessel, ℓ its length, μ the
fluid viscosity, and ΔP the pressure difference between the two ends. By applying
Poiseuille's law to the aqueous outflow, McEwen (1958) calculated that this resistance could be due to a low porosity tissue barrier averaging one 2 μm pore per .01 mm^2
surface area of Schlemm's canal. Since, according to Poiseuille's law, the resistance varies inversely as the *fourth* power of radius, a small change in the radius of
the pores can lead to a dramatic change in outflow resistance.

The outflow facility is usually calculated from the rate of change of the
intraocular volume when a tonometer is placed on the eye. The weight of the tonometer increases the pressure in the eye, causing an increased outflow of aqueous,
according to Equation (2.28), which in turn acts to reduce the intraocular pressure
(Langham, 1959a). Langham's concluding remark, however, questioned whether tonography gives a measure of the outflow resistance in the untouched eye. Bárány
(1967) gave a particularly instructive discussion of the outflow facility as it
relates to pressure changes, taking into account the reduction of aqueous secretion
at elevated pressures, which he called the "pseudofacility" in 1963. The pseudofacility is about 30% as large as the outflow facility and represents the supression
of aqueous production due to an increase in intraocular pressure. The measured
total facility is the sum of the (true) outflow facility and the pseudofacility.

The formula used to calculate the outflow facility is usually a variant of the
following expression (Saeteran, 1960c).

$$C_f = \frac{\Delta V_c}{(P_{av} - P_o - \Delta P_v)t}, \qquad\qquad (2.30)$$

where ΔV_c is the amount of corneal indentation, P_{av} the average intraocular pressure
during the time t, P_o the baseline intraocular pressure (before application of the

tonometer), and ΔP_v the rise in episcleral venous pressure with the tonometer in place, usually taken as 1.25 mmHg.

Using a constant pressure perfusion technique, Armaly (1960) found that the resistance of the outflow passages in the enucleated eye of the rabbit and cat increases linearly with the intraocular pressure. The effect of pressure change is immediate and reversible. From the graph of his results for 26 eyes, we deduce the following:

$$R_f = (0.015\ P + 0.7)R_{20}, \tag{2.31}$$

where R_{20} is the resistance to outflow at an intraocular pressure of 20 mmHg. Using the value for the outflow facility $C_f = 0.3$ μl/min/mmHg from Kornbluth and Linner (1955) for rabbit eyes and taking its reciprocal, we get $R_{20} = 3.3$ mmHg/μl/min, whereupon Equation (2.31) becomes

$$R_f = 0.05\ P + 2.3. \tag{2.32}$$

The facility of outflow is simply the reciprocal of the above equation:

$$C_f = \frac{1}{0.05\ P + 2.3}. \tag{2.33}$$

Thus, the facility of aqueous outflow is not a constant, even though most investigators consider it as such in their work. In contrast to Armaly, Levene and Hyman (1969) found in over 500 eyes that the facility of outflow *drops* 25% between two $1^1/_4$ minute intervals during tonography, where there is a continual pressure fall!

Kornbluth and Linnér (1955) performed tonography on 14 rabbits in both eyes successively and found the second eye's intraocular pressure to be 9% lower and aqueous flow 20% lower, although the facility of outflow remained the same at $C_f = 0.3$ μl/min/mmHg. The mean initial pressure in the first eye was 23.2 mmHg and in the second was 21.4 mmHg. The aqueous flow in the first was 4.4. μl/min and 3.8 μl/min in the second. The same qualitative behaviour was exhibited in humans, where they again found a constant facility of outflow of $C_f = 0.19$ μl/min/mmHg. Other values quoted in the literature for C_f for human eyes range from 0.135 to 0.33.

When the intraocular pressure is held below the venous pressure, one may study the volume flux between the intravascular and extravascular compartment across the endothelial wall of the intraocular vascular tree resulting from a balance between osmotic and hydrostatic gradients. In experiments on rhesus monkey eyes, Brubaker and Worthen (1973) evaluated the total facility, or "filtration coefficient" as 0.66 $\pm$ 0.11 μl/min/mmHg for intraocular pressure less than venous pressure. This total outflow facility for the whole eye combines three distinct components: (a) the flow across the blood-aqueous barrier due to ultrafiltration and colloid reabsorption; (b) the conventional bulk outflow; and (c) the uveoscleral flow into the suprachoroidal space.

For intraocular pressures less than the episcleral venous pressure, the hydrostatic and osmotic pressure gradients may be large enough to produce a transiently significant flux of aqueous directly across the walls of the intraocular blood vessels. In this way, an exit for aqueous is provided in response to moderate changes in the intraocular pressure without necessitating an alteration in the secretion rate to maintain equilibrium in the eye.

Pederson and Green (1973) considered the formation Q_{in} of aqueous humour by ultrafiltration Q_1 and active secretion Q_2 to be in simple equilibrium with the outflow Q_{out} by percolation Q_3 through the trabecular network and uveoscleral bulk flow Q_4:

$$Q_{in} = Q_{out}, \qquad (2.34)$$

$$Q_{in} = Q_1 + Q_2, \qquad (2.35)$$

$$Q_{out} = Q_3 + Q_4, \qquad (2.36)$$

$$Q_1 = LA(P_{cap} - P_{osm} - P), \qquad (2.37)$$

$$Q_3 = C_f(P - P_v) \qquad (2.38)$$

where L is the hydraulic conductivity of the blood-aqueous barrier, A the area available for filtration, P_{cap} the mean capillary pressure, P_{osm} the osmotic pressure difference between the blood and aqueous, P_v the episcleral venous pressure, and P the intraocular pressure. Here C_f denotes the true outflow facility, from which the effects of aqueous production have been separated.

Due to difficulties in the measurement of the four quantities P_{cap}, P_{osm}, Q_2, and Q_4, Pederson and Green (1973) proposed graphical schemes for estimating them. These methods essentially depend upon provoking changes in the intraocular and episcleral venous pressures by using sympathetic adrenergic vasoconstriction to decrease the pseudofacility, or, alternatively, upon deliberately modifying the osmotic pressure difference across the blood-aqueous barrier by known amounts.

Equation (2.34), which is strictly valid only in the absence of volume changes associated with blood flow and variation in the size of the corneo-scleral envelope, can be generalised in conjunction with the types of measurements described above, to provide estimates of these latter effects.

Eakins (1969) compared the effects of several anaesthetic agents on the aqueous humour dynamics of the cat eye and found a marked dependence upon the drug administered. For example, the outflow facility using halothane was less than half that using pentobarbital. He utilised three separate techniques to determine the facility of outflow. Both the constant infusion rate and pressure decay curve (tonography) techniques were unsuitable for repeated estimates of the outflow facility in the same eye since they yielded increasing values with each successive test. In contrast, the constant pressure method yielded reasonably stable values. He concluded that care must be taken in choosing anaesthetic agents for eye experiments and that the possible effects of the drug itself should be considered in the interpretation of the results. He seems to favour urethane as the best anaesthetic. In contrast to Eakins, Armaly (1964) found reproducible results with the constant infusion rate technique.

Finally, a further comment regarding the clinical manifestations of glaucoma may be useful. Glaucoma and raised intraocular pressure are not strictly synonymous, as glaucoma can only be diagnosed when there are changes in the visual field.

Axonal damage occurs if the blood supply to the axonal nerve fibre layer of the retina and to the optic nerve bundle is insufficient. An inadequate blood flow can result if (a) the intraocular pressure is abnormally high, causing collapse of the corresponding blood vessels (pial or retinal circulation) or (b) the intraocular pressure is normal, but a thrombosis of the central retinal vein diminishes the blood flow in the retinal circulation.

Elevated intraocular pressure is not always sufficient for impairment of the blood supply to the nerve fibres. Indeed, axonal damage may be avoided if the high intraocular pressure is accompanied by a corresponding high arterial pressure. The potentially pathological consequences of these two events, taken separately, may be effectively neutralised, since the elevated external and internal pressures to which the retinal artery is then subjected may result in an almost normal transmural pressure. Vessel collapse would then be avoided, while maintaining a normal blood supply.

The conditions defined as glaucoma must be examined in terms of their consequences for the blood supply to the nerve fibres and to the retina. Impairment of this blood supply by vessel collapse or partial occlusion will result in necrosis of certain segments of the optic nerve, with concomitant loss of part of the preprocessed visual information transmitted from the ganglion cells of the retina via the optic nerve to the brain.

Glaucomatous eyes consistently have a poorer facility of aqueous outflow, or greater resistance to outflow, than normal eyes (Grant, 1951). The rate of aqueous formation is generally somewhat lower than in the normal eye. In glaucomatous eyes, Grant (1951) found the elevation of the intraocular pressure above normal in *every* case to be due exclusively to abnormal resistance to outflow and observed no case of glaucoma due to hyperformation of aqueous. Any interference with the outflow of the aqueous humour by way of Schlemm's canal results in an immediate and marked change in the intraocular pressure.

It has been suggested by Becker and Christiansen (1956) that the empirical ratio P/C_f of intraocular pressure to outflow facility provides a useful index for differentiating between normal and glaucomatous eyes. Moreover, the magnitude of diurnal variations in intraocular pressure is increased in glaucoma, mainly through the influence of three parameters: the true outflow facility C_f, the episcleral venous pressure P_v, and the flow rate Q of aqueous humour. The total facility C_{tot} constitutes the most representative single parameter reflecting such variations

indicative of glaucoma, even though measurements of C_{tot} alone may not suffice to
identify abnormal eyes with certainty.

One may qualitatively predict the transient behaviour of the eye to a given
disturbance simply on the basis of curves of aqueous formation rates and outflow
rates plotted as functions of intraocular pressure. The ocular volume is constant
when these two rates are equal. The manner in which the eye responds to a disturbance gives an indication of the stability of the flow system. If the pressure perturbation is positive (the disturbed pressure exceeds the intraocular pressure), the
pressure will return to its normal value only if the aqueous outflow rate exceeds
the formation rate, and vice versa. Valuable clinical information concerning the
state of a particular eye is thus gained by provoking such relatively harmless disturbances, and will be elaborated upon in later sections.

5. Ocular rigidity function

As the intraocular pressure changes in response to variations in the internal
volume of the eyeball, the corneo-scleral envelope will expand and relax accordingly.
The notion of rigidity or stiffness is mainly associated with the mechanical characteristics of this envelope. This rigidity depends upon both the magnitude of the
pressure and its time rate of change, i.e. upon the viscoelastic properties of the
cornea and sclera.

Rigidity may be quantified by means of a functional relationship between intraocular pressure and volume. Various definitions have been proposed for this purpose,
in terms of "rigidity", "ocular rigidity function", or "coefficient of ocular rigidity", etc. These will be described in detail in this section, but it suffices for
the moment to note that some, such as those due to Friedenwald (1937), McBain (1958),
and McEwen *et al.* (1965), are based purely upon the elastic properties of the corneo-
scleral envelope in terms of a linear or power law dependence on the intraocular
pressure. These will lead to P-V relations which are compared graphically in
Figures 2.6 and 2.7 (pages 37 and 38). Other formulations, such as those due to
St. Helen *et al.* (1961) and ourselves, introduce a time dependence reflecting the
well-known viscoelastic properties of biological tissue. Finally, attention will be
drawn to important differences in the elastic and viscoelastic properties which may
be provoked by enucleation for the purposes of laboratory experiments.

The interior of the eye comprises solids (iris, lens, retina, vitreous, and
vascular structure) and liquids (aqueous and blood). The volume contained within
the corneo-scleral envelope in the normal human is 6000 µl (Last, 1961; Duke-Elder
and Wybar, 1961), although Ytteborg (1960d) found it to average 7700 µl. Corresponding values for cats and rabbits are 4000 µl and 1500 µl, respectively. The volumes
of solids and of vitreous vary on a slow time scale, i.e months or years. This
means that short term intraocular volume changes are due to alterations in blood or
aqueous volumes or to externally imposed volume increments, e.g. during tonometry.

The average intraocular pressure in humans is 15.5 mmHg with a normal range
from 10 to 22 mmHg. In experimental animals, normal intraocular pressures are
generally between 20 and 25 mmHg. Deep respirations can produce a rise and fall of
the intraocular pressure of as much as 5 mmHg, whereas individual pulse beats produce changes of 1 to 2 mmHg (Adler, 1965; Moses, 1970).

A. Friedenwald's formulation

The ocular rigidity relates intraocular pressure change to the corresponding
volume change. Experimentally, it is usually determined by injecting small volume
increments into the anterior chamber and measuring the resulting pressure changes
for several initial pressures.

The most commonly used pressure-volume relationship is that developed by
Friedenwald (1937), who noticed that the slope of the pressure versus volume curve
in his experiments and those of previous investigators seemed to be proportional to
the intraocular pressure:

$$\frac{dP}{dV} = aP. \tag{2.39}$$

When rearranged and integrated, this gives

$$\Delta V = \frac{1}{a}\ln(P_2/P_1) = \frac{1}{MK}\ln(P_2/P_1) = \frac{1}{K}\log(P_2/P_1), \tag{2.40}$$

where P_1 is the initial intraocular pressure and P_2 the pressure after an alteration
ΔV in the intraocular volume. (This subscript convention will be maintained in
subsequent sections.) K is called Friedenwald's coefficient of ocular rigidity.
M = 2.303 is the conversion factor between common logarithms to the base 10 and
natural logarithms to the base e. (Friedenwald's K is always calculated to the
base 10.)

Equation (2.40) is an *empirical* relation based upon thousands of experimental
observations of pressure-volume relations. Gloster (1966) reported 0.025 as the
mean value of K for humans, whereas Friedenwald gave K = 0.0215 (or a = 0.0495),
this latter figure being more generally used in the literature.

Friedenwald's equation does not represent the pressure-volume relation of the
eye with complete accuracy because in reality K is not constant. For instance, the
coefficient of ocular rigidity of enucleated human eyes decreases as the intraocular
pressure increases (Ytteborg, 1960b; Macri, Wanko and Grimes, 1958; Gloster and
Perkins, 1959). The same tendency was observed for living human eyes by Ytteborg
(1960d), who also found the rigidity coefficient is lower the greater the volume of
the eye.

The rigidity of an eye is invariably higher after enucleation than before.
Eisenlohr, Langham and Maumenee (1962) and Ytteborg (1960a) attribute this to the
retention of the entire blood volume in the enucleated eyes, whereas living eyes
have blood outflow avenues. In cats, however, enucleation seems to have no effect
on the ocular rigidity (Macri *et al.*, 1957). They found that the "rigidity" E

(defined as dP/dV) increased continuously from 15 mmHg to 30 mmHg, whereupon it stayed constant for higher pressures. Friedenwald's coefficient K also increased for rising pressures below 30 mmHg but gradually fell for higher pressures, as one might expect from a series expansion of Equation (2.40) which gives $E \simeq MKP_1$. Thus, if E is constant and P_1 rises, then K must fall. Holland, Madison and Bean (1960) found a similar behaviour for E. McEwen and St. Helen (1965), however, noticed that E continued to increase linearly with pressure in human eyes, their highest pressures being 40-50 mmHg.

B. Other empirical formulations

Because Friedenwald's equation is not valid over the full range of intraocular pressures normally encountered experimentally or clinically, numerous authors have proposed their own particular functional forms for the ocular rigidity. McBain (1958) fits his data on enucleated human eyes successfully by introducing a power relation rather than the simple linear relation used by Friedenwald:

$$\frac{dP}{dV} = aP^n, \tag{2.41}$$

which when integrated becomes

$$\Delta V = \frac{1}{a(1-n)}\{P_2^{1-n} - P_1^{1-n}\}, \tag{2.42}$$

where ΔV is the volume increment measured from the lowest pressure, P the pressure, and a and n constants. McBain's mean values for the constants are $\{a(1-n)\}^{-1} = 29.38$, and b = 1-n = 0.356. The variation in b from eye to eye is smaller than that of a, so that with a fixed value of b, one coefficient could suffice to describe a particular eye.

Holland, Madison and Bean (1960) proposed a formula similar to McBain's:

$$\frac{dP}{dV} = a(P - c)^n, \tag{2.43}$$

which integrates to

$$\Delta V = \frac{1}{a(1-n)}\{(P_2 -c)^{1-n} - (P_1 -c)^{1-n}\}. \tag{2.44}$$

The constants they found are $\{a(1-n)\}^{-1} = 2.779$, b = 1-n = 0.411, and c = 10.71. Their experiments consisted of introducing small quantities of fluid into the anterior chamber of the enucleated eye of a cat and measuring the resultant pressure changes at several constant initial pressures. They stress the importance of maintaining an enucleated eye at body temperature (37°C) during experiments, in order to minimise changes in the elastic properties of the sclera as a result of temperature variations.

McEwen and St. Helen (1965) attempted to draw together all the data in the literature for human, cat, and rabbit eyes, both living and enucleated, into a common form. They introduced the so-called *unifying formulation of ocular rigidity*:

$$\frac{dP}{dV} = aP + b, \tag{2.45}$$

which upon integration gives

$$\Delta V = \frac{1}{a}\ln \frac{P_2 + b/a}{P_1 + b/a} . \tag{2.46}$$

Friedenwald's formulation is a special case of this equation, i.e. when b = 0 and a = MK. The authors successfully applied Equation (2.46) to much of the data available in the literature. For human eyes, they give values for "a" ranging from 0.015 to 0.027 μl^{-1} and "b" from 0.03 to 0.30 mmHg/μl, with mean values for enucleated or post mortem *in situ* eyes of a = 0.022 μl^{-1} and b = 0.208 mmHg/μl.

Woo *et al.* (1972a) used a sophisticated theoretical and experimental method to determine the stress-strain properties of the sclera, stroma, and optic disc separately. They too found a linear relationship between the stress and the derivative of stress with respect to strain. In a companion paper Woo *et al.* (1972b) expressed their calculated intraocular pressure-volume relation as

$$\frac{dP}{dV} = 0.016\ P + 0.13, \tag{2.47}$$

which falls nicely within the range of values determined by McEwen and St. Helen (1965) and is virtually identical to Hibbard *et al.* (1970) for live human eyes:

$$\frac{dP}{dV} = 0.019\ P + 0.17. \tag{2.48}$$

Note that the rigidity $E \equiv dP/dV$ increases continuously with increasing intraocular pressure for each of the above four representations.

Viernstein and Cowen (1969) utilised both static and dynamic techniques to determine the pressure-volume relationship. They found somewhat different values with the two methods, although the curves were qualitatively similar. In measurements in dead rabbit eyes, they detected a markedly higher ocular rigidity than in the same eye when living, especially at elevated intraocular pressures. As reported above, enucleated human eyes display the same tendency.

More complicated finite element analyses of scleral or corneal deformation have been performed (Kobayashi *et al.* 1972a, b) in which the precise geometry, material properties, and tonometer-eye interaction are considered. These will not be discussed further here.

C. Linear elastic model

From a mathematical standpoint, the pressure-volume relationship (ocular rigidity function) is the eye's most important characteristic and is basic to the design of any mechanical experiment performed on the eye and to the interpretation of the results. In this section, we derive a relation between the intraocular pressure and the intraocular volume based on a linear elastic model.

The eye is assumed to be a perfectly elastic, thin-walled sphere (Figure 2.5). The inner radius of the sphere is R_o at zero transmural pressure. The transmural pressure for most practical purposes is simply the intraocular pressure. We assume a spatially uniform wall thickness of H_o at this state. If the intraocular pressure is raised, a circumferential strain will develop in the walls of the sphere. This strain ε_e is

$$\varepsilon_e = \frac{R - R_o}{R_o} = \frac{\Delta R}{R_o} \, , \tag{2.49}$$

where R is the radius of the sphere at the intraocular pressure P. The circumferential stress is derived by equating the forces acting on the hemisphere:

$$\sigma_e = \frac{PR}{2H} \, , \tag{2.50}$$

where H is the wall thickness in the pressurised state. Equation (2.50) assumes the stress is uniform across the entire corneo-scleral wall and is a valid approximation only for thin-walled spheres.

The volumetric strain of the human eye is small, with a maximum during an experimental test being of the order of 1%, i.e. volumetric increments of 50-60 μl in an eye of internal volume 6000 μl. For these small strains and an incompressible wall, the wall thickness in the pressurised state can be related to the wall thickness in the reference state and the respective radii:

$$H = H_o \frac{R_o^2}{R^2} \, . \tag{2.51}$$

This value for H can be inserted into Equation (2.50) to give for the stress:

$$\sigma_e = \frac{PR^3}{2H_o R_o^2} \, . \tag{2.52}$$

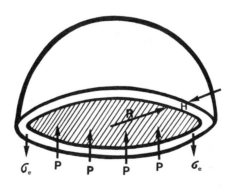

Figure 2.5. Forces acting on the corneo-scleral envelope.

The stress and strain are related by the familiar formula:

$$\sigma_e = \frac{E_e}{1 - \nu_e} \varepsilon_e ,$$

(2.53)

where E_e is the modulus of elasticity of the corneo-scleral covering of the eye and ν_e is Poisson's ratio. Equations (2.49), (2.52) and (2.53) can be combined to give:

$$P = \frac{2E_e H_o R_o \Delta R}{(1-\nu_e)R^3} .$$

(2.54)

The strain can alternatively be expressed in terms of the volume change:

$$\varepsilon_e = \frac{1}{3} \frac{V - V_o}{V_o} \equiv \frac{1}{3} \frac{\Delta V_o}{V_o} ,$$

(2.55)

where V_o is the intraocular volume at zero pressure and V the volume at the pressure P. This gives in lieu of Equation (2.54):

$$P = \frac{2E_e H_o R_o^2 \, \Delta V_o}{3(1-\nu_e) \, R^3 V_o} = \frac{2E_e H_o \, \Delta V_o}{3(1-\nu_e)R_o V} .$$

(2.56)

Because the changes in volume ΔV_o are always much less than the total volume, V can be approximated in the denominator by V_o. Then the only quantities in Equation (2.56) which do not depend upon the initial physical state of the eye are P and ΔV_o. This gives a linear relation for P and ΔV_o:

$$P = K_o \, \Delta V_o,$$

(2.57)

with K_o defined as

$$K_o = \frac{2E_e H_o}{3(1-\nu_e)R_o V_o} .$$

(2.58)

The pressure change ΔP and the volume change ΔV between any two states are then simply related:

$$\Delta P = K_o \, \Delta V.$$

(2.59)

The linearised stress analysis presented above for the corneo-scleral envelope may be refined considerably (cf. Mow, 1968) by taking into account the 5-layer sandwich structure of the corneal wall, but such an effort does not seem to be justified until better data on material properties become available.

D. Nonlinear elastic model

In a manner similar to that in Equations (2.19)-(2.22) for the arterial bed, we can consider a nonlinear elastic analysis for the corneo-scleral shell in which the incremental stress and strain are related by:

$$E_e = \frac{\text{incremental stress}}{\text{incremental strain}} = \frac{d(\frac{PR}{2H})}{dR/R} ,$$
(2.60)

with the nonlinear elasticity given by

$$E_e = E_{eo} \left(\frac{R}{R_o}\right)^{\beta} .$$
(2.61)

Combination and integration of these two equations yields:

$$P = \frac{2E_{eo}H}{\beta R} \{ \left(\frac{R}{R_o}\right)^{\beta} - 1 \} .$$
(2.62)

Substitution for H from Equation (2.51) and conversion into volumetric rather than radial values give:

$$P = \frac{2E_{eo}H_o}{\beta R_o} \left(\frac{V_o}{V}\right) \{ \left(\frac{V}{V_o}\right)^{\beta/3} - 1 \}$$
(2.63)

While the above is entirely proper mathematically, in fact, it is not essential because the radial deformation for the corneo-scleral coating is at maximum only about 1%.

In spite of the obvious temptation to simplify the mathematical formulations by applying locally measured values of the scleral rigidity to the complete outer coating of the eye, it must be noted that the distensibilities of the cornea and sclera are quite different.

Richards and Tittel (1973) pressurised the anterior chamber of enucleated human eyes by cannulation to a saline reservoir. Dimensions were recorded by a circumference gauge around the anterior-posterior and equatorial planes of the eye ball. As the intraocular pressure increased in 10 mmHg increments over 5-minute intervals up to a maximum of 120 mmHg, the distensibility $(\frac{\Delta L}{L \Delta P})$ of the sclera decreased monotonically and almost exponentially from 30×10^{-5} to 5×10^{-5} mmHg^{-1}, while corneal distensibility, which was initially lower (15×10^{-5}), attained a Gaussian-like peak of 30×10^{-5} at an intraocular pressure of about 40 mmHg, before resuming an exponential-like fall-off at values above those of the sclera. Richards and Tittel (1973) refer to the local peaking of the corneal distensibility as a "balloon" effect, indicating bulging of the cornea under pressure.

Although the cornea and sclera have very similar mechanical properties in a 3-day old infant, the greater relative distensibility of the cornea in older patients (60+ years of age) may be of importance in the design of corneal transplants. Structural changes in the stroma may lead to corneal edema and bulging at intraocular pressure of 40 to 50 mmHg. Under such conditions, the endothelium no longer covers the posterior surface and separation between the cells can lead to an accumulation of fluid, as seen clinically in patients with rapid increases in intraocular pressures to these levels.

Graebel and van Alphen (1977) recently showed the rate of relaxation of the sclera is conditioned by the contiguous choroid, although the latter has little structural strength in itself. In fact, the sclera is 5 times thicker than the choroid, with an elastic modulus one order of magnitude larger. Uniaxial stress-strain measurements at constant strain rate performed on isolated strips of sclera and choroid from human eyes indicate an exponential relation for the sclera

$$\sigma_{scl} = A \, (e^{a\varepsilon}-1), \tag{2.64}$$

similar to many other biological soft tissues (Fung, 1967), and a power law behaviour for the choroid

$$\sigma_{ch} = \sigma_0 \, (\varepsilon/\varepsilon_0)^{\alpha}, \tag{2.65}$$

where A, a, and α are constants, and the subscript "o" denotes reference conditions. There is a great deal of scatter in Graebel and van Alphen's results.

The choroidal properties appear to play an important role in the accommodation of the lens, as the choroidal elasticity counteracts the force of the ciliary muscles and zonule fibre-lens combination (Moses, 1971). The pull on the choroid is least when the ciliary muscle is relaxed and the lens is under the greatest tension, and vice versa when the ciliary muscle is contracted. Stiffening of the choroid with age may inhibit proper focussing of the lens and lead to presbyopia, i.e. the near point of distinct vision is farther from the eye than normal.

A completely satisfactory nonlinear elastic model to explain the empirical ocular rigidity function does not yet exist.

E. Comparison of ocular rigidity functions

It is difficult to make a simple comparison of the five ocular rigidity functions considered so far, as the parametric values given by different authors are necessarily derived from different experiments and, in the case of Holland, Madison and Bean (1960), from different species. But the primary obstacle to direct comparison lies in the functional forms assumed for the pressure-volume relation: the linear elastic and Friedenwald formulations have one parameter; McBain's and McEwen and St. Helen's have two; and Holland, Madison and Bean's has three. Clearly, the more parameters in an expression, the more accurately it should fit experimental data.

The comparative approach we adopt here is to require all formulations to have the same pressure-volume slope at $(\Delta V, P) = (0, 15.5)$ as Friedenwald's formulation: slope $\equiv dP/dV = 0.767$ mmHg/μl. For formulations which have more than one parameter, we have adjusted only the initial factor "a" and left the others as given by the authors. The formulas obtained in this manner are

Friedenwald:
$$\Delta V = 20.2 \ln (P_2/P_1) \tag{2.66}$$

37

McBain: $\qquad\qquad\qquad\qquad\Delta V = 21.4 \{P_2^{.356} - P_1^{.356}\}$ (2.67)

Holland, Madison & Bean: $\quad \Delta V = 3.93\{(P_2 - 10.7)^{.611} - (P_1 - 10.7)^{.611}\}$ (2.68)

McEwen & St. Helen: $\qquad\quad \Delta V = 27.2 \ln \frac{P_2 + 5.8}{P_1 + 5.8}$ (2.69)

Linear: $\qquad\qquad\qquad\qquad \Delta V = 1.304 (P_2 - P_1)$ (2.70)

where P_1 = 15.5 mmHg. These formulas are plotted in Figure 2.6, along with experimental values for McBain's enucleated human eye No. 106, whose constants clearly differ from those given in Equation (2.67).

Alternatively, the formulas can be plotted as in Figure 2.7 using the authors' own constants, which give for

McBain: $\qquad\qquad\qquad\qquad \Delta V = 29.38\{P_2^{.356} - P_1^{.356}\}$ (2.71)

Holland, Madison & Bean: $\quad \Delta V = 2.779\{(P_2 - 10.7)^{.611} - (P_1 - 10.7)^{.611}\}$ (2.72)

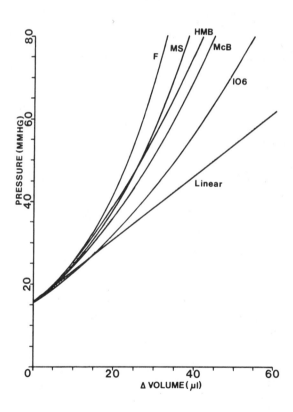

Figure 2.6. Comparison of ocular rigidity functions with identical slopes at P = 15.5 mmHg. F = Friedenwald; MS = McEwen & St. Helen; HMB = Holland, Madison & Bean; McB = McBain; 106 = McBain's experimental result.

38

McEwen & St. Helen: $\qquad \Delta V = 45.45 \ln \dfrac{P_2 + 9.45}{P_1 + 9.45}$ (2.73)

The Friedenwald and linear elastic formulas are as before. The experimental values
for McBain's eye No. 106 are again given for comparison. The values for Holland
et al. are for cats, whereas the other four are for humans. Note that Friedenwald's
formula gives results for humans considerably different from the others.

F. Viscoelasticity of corneo-scleral envelope

St. Helen and McEwen (1961) demonstrated the viscoelastic behaviour of human
sclera, i.e. the time dependency of the ocular rigidity. By injecting volume
increments into scleral segments and noting the resultant pressure decay while the
sclera was kept at constant strain, they observed that the pressure curve comprises
an equilibrium pressure difference, a fast relaxation curve, and a slow relaxation
curve:

$$\Delta P = \Delta P_{eq} + \Delta P_{fast} + \Delta P_{slow} = \Delta V(K_{eq} + K_f e^{-m_f t} + K_s e^{-m_s t}), \qquad (2.74)$$

where the K's are *new* rigidity constants, and the m's time constants. Equation
(2.74) may be considered a viscoelastic generalisation of the elastic model (2.45)

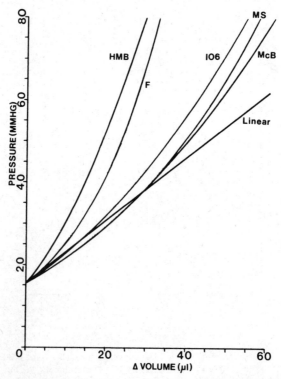

Figure 2.7. Comparison of ocular rigidity functions with constants as given by the
authors. Abbreviations as for Figure 2.6.

of McEwen and St. Helen (1965) for t = 0, provided the sum of the rigidity constants equals aP + b.

The half-lives of the decaying exponentials were determined by St. Helen and McEwen in two sets of experiments, lasting four minutes and thirty minutes, respectively. The following are their average 30-minute values for positive ΔV and initial pressure = 12 mmHg, which we believe are more reliable than the 4-minute values:

$$K_{eq} = 0.5 \text{ mmHg/}\mu l, \quad K_f = 1.3 \text{ mmHg/}\mu l, \quad K_s = 1.6 \text{ mmHg/}\mu l$$
$$m_f = 1.82 \text{ min}^{-1}, \quad m_s = 0.072 \text{ min}^{-1}.$$

The 30-minute values for negative ΔV were considerably different:

$$K_{eq} = 1.4 \text{ mmHg/}\mu l, \quad K_f = 0.5 \text{ mmHg/}\mu l, \quad K_s = 0.3 \text{ mmHg/}\mu l$$
$$m_f = 2.31 \text{ min}^{-1}, \quad m_s = 0.147 \text{ min}^{-1}.$$

As we will be concerned primarily with elevated pressures, we will adopt the 30-minute values corresponding to positive ΔV, although in a later paper, McEwen, Shepherd, and McBain (1967) used the 4-minute values. The essential point, however, is that the sclera is a viscoelastic material.

The viscoelastic effects in enucleated human eyes have also been investigated by Schlegel *et al.* (1972). Two independent methods were used to measure the instantaneous dimensions of the enucleated eye: (a) an ultrasound phase shift, with an estimated resolution of 1 μm, and (b) a "flying-spot" micrometer consisting of a photomultiplier recording of a sweeping electron beam as it reflects from two white targets on the eye segment, with an estimated resolution of 5 μm. The results from both methods were consistent and indicated that the externally-measured viscoelastic response in enucleated human eyes decreases rapidly over 8 hours for a given pressure change. The stored eyes were subjected to 42-minute loading cycles alternating between 20 and 40 mmHg. Viscoelastic effects were found to be greater in the earlier part of loading experiments and were accentuated for eyes stored at lower pressures in comparison with eyes maintained at 15 mmHg by a saline reservoir.

The viscoelastic response, which may be quantified by a sum of terms decreasing exponentially in time as done by St. Helen and McEwen (1961), is large enough to cause significant errors in measurement of pressure-volume and pressure-flow relationships, unless appropriate corrections are introduced. For example, the viscoelastic effect of the outer coats of the eye may lead to an error of 5 μl/min in the flow rate during the first hour of loading for a change in pressure of 20 mmHg.

Because the sclera is not of constant thickness, the rigidity of scleral segments will vary according to the particular area chosen for the sample (Hibbard *et al.*, 1970; van der Werff, 1972). The behaviour of the whole eye must be considered in any determination of the rigidity. Therefore, we assume that the K's found by St. Helen and McEwen maintain their same *relative* magnitudes for the entire scleral-corneal membrane, but that K_t defined as

$$K_t \equiv K_{eq} + K_f + K_s \tag{2.75}$$

must be determined from measurements on complete eyes. Equation (2.74) can then be rewritten:

$$P - P_o = K_t \, \Delta V(K_{eq}^* + K_f^* \, m^{-m}f^t + K_s^* \, e^{-m}s^t). \tag{2.76}$$

where $K_{eq}^* = K_{eq}/K_t$, etc.

During any *rapid* stress/strain experiment, relaxation effects would not occur: the eye would behave as if the ocular rigidity constant were simply K_t. The same would apply to pressure pulsations during the cardiac and respiratory cycles. If the pressure excursions during the experiment are large enough, however, the rigidity K_t might have to be replaced by a nonlinear, pressure-dependent function.

6. Causal relationships (influence diagram)

Drawing upon the observations made in the preceding sections, causal relationships between the various parameters of the eye can be postulated, as shown in Figure 2.8. The arrows denote the direction of the causal relationship, from cause to effect, with solid lines denoting positive influence and dashed lines negative influence. By positive (negative) influence is meant a numerically positive change in one quantity causes a numerically positive (negative) response in

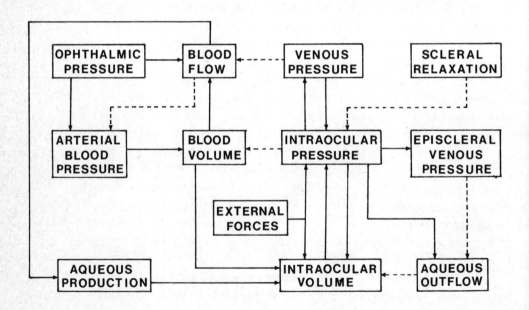

Figure 2.8. Influence diagram.

the influenced quantity. A similar diagram with five parameters was published by
Bill (1962a).

Basic to the eye's mechanical performance is the pressure-volume relationship,
i.e. the ocular rigidity function. If the pressure is increased, so is the volume,
and conversely. Thus, positive arrows are drawn from each to the other.

The blood volume is directly related to the transmural pressure, i.e. the
intraocular arterial pressure minus the intraocular pressure. A rise in the arterial pressure increases the blood volume, whereas a rise in the intraocular pressure
decreases it. The blood volume is a component of the total intraocular volume and,
thus, influences it positively.

The intraocular arterial (retinal or choroidal) pressure primarily reflects
the ophthalmic artery pressure. It is negatively influenced, however, by the blood
flow rate through it, as can be seen from a consideration of Poiseuille's law which
relates the flow through a tube linearly to the pressure drop along it. If the flow
is increased, so is the pressure drop. But a larger pressure drop means a lower
absolute pressure. Hence, the negative influence.

In much the same way, the blood flow is influenced by the two pressures which
drive it. The ophthalmic (higher) pressure influences the blood flow positively,
while the venous (lower) pressure influences it negatively. The resistance
encountered by the blood flowing through the eye is essentially a function of the
local radii of the blood vessels which is equivalently represented by the total
blood volume in the eye. A higher blood volume means a lower resistance which in
turn increases the blood flow for a given perfusion pressure.

The venous pressure affects and is affected by the intraocular pressure. A
rise in one induces a rise in the other, primarily due to the poor distensibility
of the veins.

A rise in the rate of production of aqueous humour increases the volume of the
eye. A decreased blood flow, on the other hand lowers this rate of production.

The rate of aqueous outflow affects the intraocular volume in the opposite
manner, i.e. a rise in the outflow decreases the intraocular volume. The difference
between the intraocular and episcleral venous pressures drives the aqueous outflow
in the same way that the difference between the ophthalmic and venous pressures
drives the blood flow. The episcleral pressure is directly and positively influenced by the intraocular pressure.

Scleral relaxation refers to the gradual reduction in the scleral stress when
the intraocular volume is maintained at an elevated level for an extended period of
time as in the experiments by St. Helen and McEwen (1961). It is shown as a
negative influence because the pressure falls with time.

Finally, we include the (generally) positive influence of external forces such
as obtained with tonometry, ophthalmodynamometry, suction cups, or stray fingers.
These are shown as double arrows to both the intraocular volume and the intraocular

pressure. Which of these two affected quantities is chosen generally depends upon the analyst's personal preference although the choice is frequently dictated by the experimental design. Other parameters in the influence diagram could also be varied by external forces, but for simplicity's sake, these are not shown.

Figure 2.8 pictorially demonstrates the causal relations found in the eye and gives a quick visual reinforcement of the influence patterns. By disturbing any of the parameters and tracing the effects through the diagram, the homeostatic nature of the eye's behaviour can be seen.

7. The Standard Eye

The "Standard Eye" is defined to be a hypothetical human eye whose parameters, in most cases, correspond to the best experimental values obtainable from the literature. These values are taken from a variety of sources, so there is no *a priori* reason for all of them to be necessarily consistent one with another, although we have tried to make them so.

The value of the Standard Eye lies in providing a suitable basis of comparison when parameters are varied or new models developed. The Standard Eye is *not* necessarily the one which gives the best predictions for a particular situation.

The Standard Eye's parameters are

Intraocular volume	6000 µl
Total eye volume	7000 µl
Anterior chamber volume	250 µl
Posterior chamber volume	60 µl
Mean intraocular pressure	15.5 mmHg
Uveal blood pressure	
Systolic	75 mmHg
Diastolic	35 mmHg
Intraocular blood volume	18 µl
Retinal circulation time	1.3 sec
Intraocular blood flow rate	0.85 ml/min
Aqueous volume flow rate	2.0 µl/min
Outflow facility	0.19 µl/min/mmHg
Ocular rigidity at mean intraocular pressure	0.767 mmHg/µl
McEwen & St. Helen's parameter "a"	$0.022 \ \mu l^{-1}$
McEwen & St. Helen's parameter "b"	0.208 mmHg/µl
St. Helen & McEwen's parameter K_{eq}^{*}	0.15
St. Helen & McEwen's parameter K_{f}^{*}	0.38
St. Helen & McEwen's parameter K_{s}^{*}	0.47

8. A summary of normal values and relations

It is clear from the preceding sections that data in the literature refer to different species. In order to help the reader, we have prepared the following summary tables of the important parameters for humans, cats, and rabbits, the latter two being the principal experimental animals. Most of the values presented refer to averages of many experiments. We would, thus, repeat our earlier caution against uncritical use of these parameters for a particular individual.

TABLE II.   AQUEOUS CHARACTERISTICS

Flow rate (μl/min

Q_{aq} = 2.4 Grant (1951)
  = 2.2 Schimek (1964)
  = 2.75 Kinsey *et al.* (1949)
  $\left.\begin{array}{l} = 3.82 \\ = 4.37 \end{array}\right\}$ Kornbluth & Linnér (1955)
  = 1.61 in monkeys Bill (1969)

Volume turnover (/min)

$\dfrac{\Delta V_{aq}}{\Delta t}$ = 1.9% Goldmann (1955)

  = 1.1 - 1.9% in rabbits Moses (1970)

  = 20%/min of total Vol_{ah}

    → 50 μl/min in rabbits
         Kinsey *et al.* (1949)
  = 5.2-15.6% Moses (1970)

Outflow facility (μl/mm/mmHg)

C_f = 0.19
  = 0.66 ± 0.11 in monkeys
       Brubaker *et al.* (1973)
  = 0.3 in rabbits
       Kornbluth *et al.* (1955)

role of anaesthetic agents:
    Eakins (1969)

$C_{f_{halothane}} < \frac{1}{2} C_{f_{pentobarbital}}$

$C_{f_{hyaluronidase}} > C_{f_{normal}}$ Grant (1958)

Viscosity (centipoise)
  μ = 1.03 - 1.04 Moses (1970)

TABLE III. VASCULAR CHARACTERISTICS

	Retina	Choroid	Uvea	Episclera	Eye
Pressure (mmHg)	70/35 (Magitot et al. 1922) 75/35 (Cole 1966) $P_{syst}=P_{oph} + (14\text{-}17)$ (Bill 1963a; van der Werff 1971) $P_{ret}-P_{brach}= 20.4$ diast. (Perry & Rose 1958) $P_{oph}-P_{brach}= 5.3$ (Borrás et al. 1969b)	$P_{choroidal} = P - 2.7$ venous (Macri 1964)		$P_v = P + 2$ (Duke-Elder 1926) $P_v=8.94$ in rabbits (Kornbluth et al. 1955)	$P = 15.5$ $P=P_{ant.cil.} -0.2$ $P=P_{CH} + 5.9$ (Macri 1964)
Blood flow rate (ml/min)	$Q = 1.72/gm$ retina (Anderson & Saltzmann 1964)	$Q=50/(gm$ choroid) $Q=10/(gm$ chor. + ret.) (Bill 1967) $Q=16.6/(gm$ choroid) in rabbits (Wilson et al. 1973)	$Q=2\text{-}2.4$ in rabbits $Q=1.2\text{-}1.5$ in cats (Bill 1962a,b; 1967; 1970)		$Q=1.30$ (Aronson 1974) $Q=0.85$ in rabbits (Levene 1957)
Blood velocity (mm/sec)	$v_{artery}= 22$ (100 μm dia) $v_{vein} = 19$ (160 μm dia) $v_{vein} = 16$ (130 μm dia) (Tanaka et al. 1974)				
Retinal circulation time (sec)	$RCT = 1.3$ (Suvanto et al. 1960) $RCT = 4.7 \pm 1.1$ (Hickman et al. 1965) $RCT = \dfrac{V}{Q} = \dfrac{6.8}{5.3} = 1.3$				
Blood volume (μl)			$\dfrac{V_{uveal}}{V_{ret}} = 37$ in cats (Chao & Bettman 1957)		$V_{eye} = 18.2$ /eye $\to 6.8$ /gm eye (Fish et al. 1969)

CHAPTER 3. GENERAL TIME-DEPENDENT MODEL

Weinbaum (1965) proposed a simplified mathematical model for the human eye which constitutes a valuable framework. Although his model provides for changes in the volume of the vascular bed and in the aqueous humour formation rate, the pressure-dependence of the latter was not then available. In order to obtain an analytical solution, he found it necessary to consider only *linear* viscoelasticity, i.e. just one but not both of the fast and slow relaxation exponentials.

McEwen, Shepherd and McBain wrote three papers in 1967 on a general electrical analogue model of the human eye. The first paper describes the fundamental equations and the electrical circuitry for the model. For the behaviour of the scleral wall, they employed St. Helen and McEwen's time-dependent stress relaxation (our Equation (2.74)). In the second and third papers (Shepherd *et al.*, 1967; McBain *et al.*, 1967) this model was used to compute the pressures for the tonographic and suction cup procedures. The results agree well with clinical recordings.

We devote this chapter to incorporating the separate elemental volume changes discussed in the previous chapter into a single mathematical statement for the volume equilibrium of the eye. This is transformed into a differential equation governing the time rate of change of the intraocular pressure. Analytical and numerical solutions are then obtained from the general equation.

In setting up the differential equations for the behaviour of the human eye, we shall take into account all of the relationships discussed in Chapter 2 and summarised in the influence diagram (Figure 2.8).

1. Governing equations

The intraocular volume (V) responds to changes in both aqueous (V_{aq}) and blood (V_a) volumes, in addition to external volume changes imposed (V_{ext}), e.g. by tonometric probes on the cornea. The time rates of change of volume are simply related:

$$\frac{dV}{dt} = \frac{dV_{aq}}{dt} + \frac{dV_a}{dt} + \frac{dV_{ext}}{dt} . \tag{3.1}$$

The rate of change of aqueous volume is the difference between the aqueous production and outflow rates:

$$\frac{dV_{aq}}{dt} = S_p - S_o. \tag{3.2}$$

Previously, we derived the dependence of S_p and S_o on pressure. We found that the aqueous production decreases linearly with intraocular pressure and stops altogether when a critical pressure is reached:

$$S_p = C_p(P_c - P), \tag{2.24}$$

that the aqueous outflow is proportional to the difference between the intraocular and episcleral venous pressures:

$$S_o = C_f(P - P_v),$$ (2.28)

and that the outflow facility is pressure-dependent:

$$C_f = \frac{1}{a_1 P + a_2},$$ (3.3)

with the estimated values for rabbit eyes being $a_1 = 0.05$ and $a_2 = 2.3$ as given in Equation (2.32). Assuming this value for a_1 also holds for human eyes, consistency requires that a_2 be adjusted to 1.125.

We saw earlier that the episcleral venous pressure itself is a function of the intraocular pressure, of the form:

$$P_v = a_3 P + a_4.$$ (3.4)

The values of $a_3 = 0.48$ and $a_4 = 3.1$ as given in Equation (2.1) are not compatible with the respective mean intraocular and episcleral venous pressures of 15.5 mmHg and 11.7 mmHg. Assuming that a_3 is correct, a value of $a_4 = 4.26$ leads to compatible values for the mean pressures. Incorporating the above equations into Equation (3.2) yields:

$$\frac{dV_{aq}}{dt} = C_p(P_c - P) - \frac{(1-a_3)P-a_4}{a_1 P + a_2}.$$ (3.5)

We must, of course, remember that the constants appearing in the above equation vary from eye to eye. Nevertheless, applying these constants to the Standard Eye we obtain:

$$\frac{dV_{aq}}{dt} = 0.027(90 - P) - \frac{0.52P - 4.26}{0.05P + 1.125}.$$ (3.6)

The pressure-volume relation for the vascular bed given in Equation (2.18) can be differentiated with respect to time to yield:

$$\frac{dV_a}{dt} = \frac{V_a}{K_\alpha(P_a-P)^\alpha} \left(\frac{dP_a}{dt} - \frac{dP}{dt}\right).$$ (3.7)

We now consider the important intraocular pressure-volume relationship. In the preceding Chapter, five different ocular rigidity functions were presented, none of which was time-dependent. Yet, Equation (2.74) implies ocular rigidity _is_ time-dependent, i.e. the corneo-scleral shell is viscoelastic. In order to resolve this paradox, let us write Equation (2.76) in its differential form:

$$\frac{dP}{dV} = K_t(K_{eq}^* + K_f^* e^{-m_f t} + K_s^* e^{-m_s t}),$$ (3.8)

and consider its behaviour in the limit for times $t \ll 1$. The terms in brackets sum identically to 1 and the right hand side of Equation (3.8) reduces simply to K_t.

If K_t were a constant, the ocular rigidity dP/dV would be the same as given in the linear elastic analysis, Equation (2.59). All of the experimental evidence, however, indicates that the ocular rigidity is *not* constant, but pressure-dependent. We can thus identify K_t with one of the pressure-dependent ocular rigidity functions discussed in the preceding Chapter. For this purpose, we prefer the precision of McEwen and St. Helen's formulation as given in Equation (2.45). Equation (3.8) above can now be rewritten as

$$\frac{dP}{dV} = (aP + b) (K^*_{eq} + K^*_f e^{-m_f t} + K^*_s e^{-m_s t}),$$ (3.9)

which incorporates the desired dependence of the ocular rigidity on both pressure and time.

As an aside, we note many investigators wait a short time for the pressure to stabilise after a volume increment has been introduced into the eye. We suggest that these investigators may be losing most, if not all, of the "fast" component $K^*_f \exp(-m_f t)$ of the ocular rigidity. They would, thus, report an ocular rigidity about two-thirds of its true value.

The intraocular volume changes can be expressed in terms of pressure changes, since clinically or experimentally it is easier to measure pressures than volumes:

$$\frac{dV}{dt} = \frac{dV}{dP}\frac{dP}{dt} = \frac{1}{dP/dV}\frac{dP}{dt} .$$ (3.10)

Equations (3.1), (3.5), (3.7), (3.9), and (3.10) can be combined to give the final form of the general governing equation for intraocular pressure:

$$\frac{dP}{dt} = \frac{C_p(P_c - P) - \dfrac{(1-a_3)P - a_4}{a_1 P + a_2} + \dfrac{V_a}{K_\alpha (P_a - P)^\alpha}\dfrac{dP_a}{dt} + \dfrac{dV_{ext}}{dt}}{\dfrac{1}{(aP + b)(K^*_{eq} + K^*_f e^{-m_f t} + K_s e^{-m_s t})} + \dfrac{V_a}{K_\alpha (P_a - P)^\alpha}} .$$ (3.11)

Equation (3.11) defines the time variation of the intraocular pressure in terms of the rigidities of the corneo-scleral envelope and the blood vessels, the production and outflow of aqueous humour, the arterial pressure, and external volume changes, for given values of the intraocular pressure.

It is sometimes convenient to express external influences in terms of pressure rather than volume, in which case the last term in the numerator of Equation (3.11) would be replaced by

$$\frac{dV_{ext}}{dt} = \frac{dV_{ext}}{dP}\frac{dP}{dt} .$$ (3.12)

2. Approximate analytical solutions

The difficulty in many complex mathematical analyses lies less in formulating the general governing equations than in solving them. Such is our difficulty as well. Thus, recourse is made to approximations so that we can obtain some useful analytical results and apply them practically to common situations involving the eye.

Equation (3.11) can be broken into two parts: one governing the mean pressure $\bar{P}$; the other governing the pulsations P^* which result from variations in the arterial pressure. The pressure at any time is the sum of these two components:

$$P = \bar{P} + P^*. \tag{3.13}$$

(The mean pressure $\bar{P}$ is not to be confused with the steady-state intraocular pressure P_o = 15.5 mmHg. Rather, $\bar{P}$ is the time-dependent pressure obtained when second-by-second variations are averaged out.)

The cardiac period of approximately one second is much shorter than the two characteristic relaxation times so that a quasi-steady approximation can be made for the pressure pulsations. Specifically, we denote by K_T the value of the viscoelastic terms at time T, i.e.

$$K_T = (aP + b)(K_{eq}^* + K_f^* e^{-m_f T} + K_s^* e^{-m_s T}). \tag{3.14}$$

During a typical pulsation, K_T can be considered time-independent. It has a maximum value of $(aP + b)$ and a minimum of $(aP + b) K_{eq}^*$. Substituting Equations (3.13) and (3.14) into (3.11) and assuming $P^* \ll \bar{P}$, the two equations governing P^* and $\bar{P}$ can be identified:

$$\frac{dP^*}{dt} = \frac{1}{\dfrac{K_\alpha (P_a - \bar{P})^\alpha}{V_a K_T} + 1} \frac{dP_a}{dt}, \tag{3.15}$$

$$\frac{d\bar{P}}{dt} = \frac{C_p(P_c - \bar{P}) - \dfrac{(1-a_3)\bar{P} - a_4}{a_1 \bar{P} + a_2} + \dfrac{dV_{ext}}{dt}}{\dfrac{1}{(a\bar{P} + b)(K_{eq}^* + K_f^* e^{-m_f t} + K_s^* e^{-m_s t})} + \dfrac{V_a}{K_\alpha (P_a - \bar{P})^\alpha}}. \tag{3.16}$$

A. Steady state pressure pulsations

The simplest problem to which Equation (3.15) can be applied is to the determination of the intraocular pressure pulsations about the mean steady state pressure $(\bar{P} = P_o)$ as a function of arterial pressure pulsations. For this case, there are no external forces on the eye, the aqueous production and outflow rates are essentially constant and equal, and the time period of the pulsation is short enough that the viscoelastic behaviour of the sclera does not come into play. Under these conditions, Equation (3.15) reduces to

$$\frac{dP^*}{dt} = \frac{1}{\dfrac{K_\alpha(P_a-P_o)^\alpha}{V_{ao}K_T} + 1} \frac{dP_a}{dt} \; , \tag{3.17}$$

where $K_T = aP_o + b$ and V_a has been given its steady state value V_{ao}.

The difficulty is that the factor in this equation is by no means constant since P_a-P_o varies widely. Cole (1966) reports the systolic and diastolic pressures in the uvea as 75/35, i.e. an arterial pressure pulsation of 40 mmHg. With P_o = 15.5 mmHg, this gives a maximum value of P_a-P_o of about 60 mmHg and minimum of about 20 mmHg.

Equation (3.17) can be integrated exactly only for rational values of α. Fortunately, α = 1.6 is a rational value, i.e. 8/5. The values for K_α, V_{ao}, and K_T were found previously in Chapter 2. Integrating Equation (3.17) we find that for arterial pressure pulsations of 40 mmHg the corresponding intraocular pressure pulsations are 2.7 mmHg. Considering the approximations made, these predicted peak-to-peak pulsations compare well with the observed values of 1 to 2 mmHg (Adler, 1970).

We can estimate the blood volume pulse by means of the vascular rigidity of Equation (2.23) to give ΔV_a = 1.2 µl.

B. Non-steady state pressure pulsations

When the intraocular pressure is not at its steady state value ($\bar{P} \neq P_o$), some of the assumptions of the preceding section are no longer valid. Specifically, the aqueous production will no longer equal the outflow, nor will the blood volume equal its steady state value. The assumption of small intraocular pressure pulsations is still valid, however, and Equation (3.15) continues to govern this situation.

The difficulties experienced in the preceding section in integrating Equation (3.15) are compounded here by the nonlinear character of the factor $K_\alpha(P_a-\bar{P})^\alpha/V_a$. In addition, the possibility that the vascular bed may collapse during the diastolic portion of the cardiac cycle cannot be excluded, in which case the model itself would be invalid.

At the beginning of tonography, for example, the initial intraocular pressure is higher than the steady state pressure P_o. Comparison of Equation (3.15) and (3.17) then leads one to conclude that the pressure pulsations will be larger than at steady state. These will decrease gradually as both the mean pressure falls and viscoelastic relaxation becomes appreciable.

3. Numerical solutions

Although the foregoing approximations permit analytical solutions which provide insight into the mechanical response of the eye, it becomes necessary to return to the general governing Equation (3.11) to obtain quantitative predictions suitable for comparison with clinical measurements.

At this point in the development of the model, we must rely heavily upon empirical data which are unfortunately still sparse and uncertain. Much has been reported in the literature on ocular measurements on cats, rabbits and humans, as has been covered in Chapter 2, but controversy still remains.

In order to illustrate the comprehensiveness of the mathematical model, we apply it to a specific eye, using parameters which might be encountered clinically. This case represents the ocular dynamics of a normal eye with consideration of the mechanical regulatory mechanisms only.

The pressure within the blood vessels of the eye is chosen to have an average value of P_a = 73 mmHg, since beat to beat variations are not of particular interest to us here.

Equation (3.16) can be integrated numerically to yield the time variation of the mean intraocular pressure subject to the following initial conditions, i.e. at time t = 0:

$$P_o = 15.5 \text{ mmHg}$$
$$P_a = 73 \text{ mmHg (constant)}$$
$$\bar{P} = 25 \text{ mmHg}$$

$$\frac{dV_{ext}}{dt} = 0$$

with the following values for the aqueous humour parameters:

a_1 = 0.0306 $(\mu l/min)^{-1}$ a_2 = 1.428 mmHg/μl/min

a_3 = 0.5 (dimensionless) a_4 = 3.95 mmHg

C_p = 0.046 μl/min/mmHg P_c = 90 mmHg

and with the following ocular and blood vessel rigidities:

K_{eq}^{*} = 0.15 K_f^{*} = 0.38 K_s^{*} = 0.47

$K_t = a\bar{P} + b$ = 0.53 mmHg/μl m_f = 1.803 min^{-1} m_s = 0.072 min^{-1}

a = 0.0213 μl^{-1} b = 0.0 $\dfrac{K_\alpha (P_a - P)^\alpha}{V_a}$ = 9.39 mmHg/μl

The results of a parametric study of the influence on the intraocular pressure due of variations in the outflow facility parameter (a_2), the ocular rigidity (K_t), and aqueous cutoff pressure (P_c) for this particular eye are presented in Figures (3.1), (3.2), and (3.3), respectively. The following discussion is intended to illustrate the sensitivity of the intraocular pressure to changes in the above parameters, without consideration of the underlying physiological causes or consequences, a full investigation of which lies outside the scope of this preliminary study.

The curves indicate modest changes in the intraocular pressure for modest variations of the three parameters a_2, K_t, and P_c.

In accordance with Equation (3.3), a rise in a_2 leads to a drop in C_f, the aqueous outflow facility, and hence by Equation (2.28) to a drop in the outflow S_o. If the outflow is inhibited, the intraocular pressure should rise, as demonstrated by the curves in Figure 3.1.

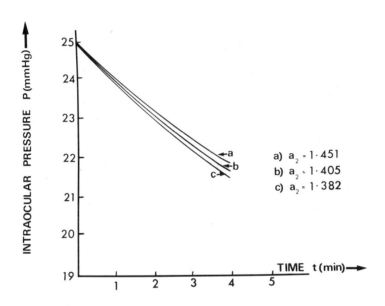

Figure 3.1. Effect of variations in the aqueous outflow parameter a_2 on the intra-ocular pressure.

Since the ocular rigidity dP/dV increases with increasing K_t, cf. Equation (3.8), pressure changes become larger for a given volume increment, i.e. the corneo-scleral envelope becomes more rigid, and the intraocular pressure decreases more rapidly (under the relieving action of the aqueous outflow), as shown in Figure 3.2.

The rate of aqueous secretion into the eye is affected by changes in the cutoff pressure P_c. A rise in the threshold P_c leads to an increase in the form-ation rate S_p {Equation (2.24)}. This implies a higher rate of accumulation of aqueous in the eye and an increase in the ocular pressure, as borne out by the curves of Figure 3.3.

Although plausible in the neighbourhood of the normal physiological values, these sample results fail to account for all aspects of observed behaviour in the

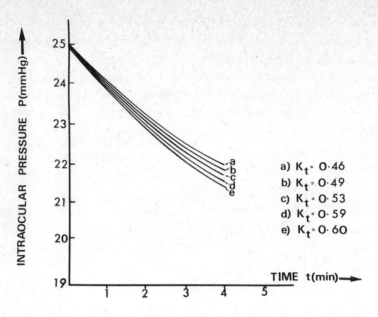

Figure 3.2. Effect of variations in the ocular rigidity K_t on the intraocular pressure.

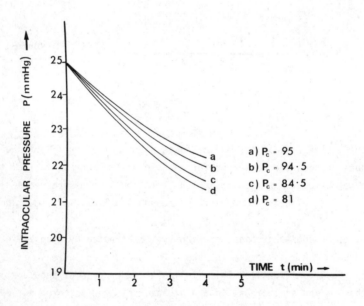

Figure 3.3. Effect of variations in the cutoff pressure P_c on the intraocular pressure.

pathological range. For example, Gloster (1966) reports a sharp rise in the outflow facility C_f for low values of the ocular rigidity K_t lying between 0.1 and 0.4. Now, for each integrated value of the intraocular pressure $\bar{P}$ obtained from Equation (3.16), there corresponds a value C_f of the outflow facility from Equation (3.3). However, the present model would indicate only a small variation in C_f with the values used in the above example. This implies that the simplified theory either does not contain all the dominant control features or the functional forms introduced are lacking in generality. Further research is required in this direction.

CHAPTER 4. NEURAL CONTROL OF THE INTRAOCULAR PRESSURE

Neural control of intraocular parameters has not been considered in the above computations, and this may explain the inadequacy of the model's predictions in the pathological range. Little is known about the neural control mechanisms in the eye, although drugs are administered in the clinical treatment of glaucoma with varying degrees of success. Such drugs may act either directly on the eye or secondarily through altered neural control. The neural pathways controlling the intraocular pressure are largely unknown, and one must assume them in a preliminary study.

The intraocular pressure is determined mainly by the formation and outflow rates of aqueous humour. The problem in establishing a control model is two-fold: first, the location of the neural sensors must be assumed; and second, their mode of action on the flow system must be described.

Concerning the location, Hart's (1970) histological studies indicated the presence of a network with bare axonal endings associated with the trabeculae. That these endings are undifferentiated may indicate that they respond to 'squeezing' during deformation of the surrounding tissue. Quantitatively, the "trabecular nerves" would be sensitive to the pressure gradient, or equivalently, to the product of flow rate and trabecular resistance. Hart also found the eyeball to contain neurons likely to be sensitive to wall strain.

As for the mode of action of these sensors, there is no histological evidence that the dimensions of the aqueous outflow channels are controllable. Hart suggested that the flow resistance is influenced by vasomotor control of the arterial resistance supplying the major circle of the iris (Figure 4.1). That is,

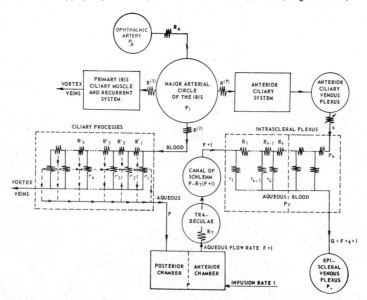

Figure 4.1. Arterial vasomotor control in the iris. From Hart (1972) - used by permission.

the trabecular and scleral wall nerve 'sensors' are linked to vasoconstrictive and vasodilative control of the arterial lumen, in such a way that both the formation of aqueous by blood flow through the ciliary processes and the aqueous outflow in the drainage canals shared by both blood and aqueous are modified. For example, it could be postulated that arterial conductance is controlled linearly by vasocontriction excited by the wall strain sensors, while vasodilation is triggered by the flow-resistance product sensors. The balance of the two competing vasomotor controls would then determine the aqueous flow equilibrium and, hence, the intraocular pressure.

The rate of formation of aqueous humour, which according to Equation (2.24) would passively decrease with increasing intraocular pressure, could, under neural regulation, be maintained at a nearly constant value over a substantial range of the intraocular pressure. Indeed, there is some evidence which indicates that it is the aqueous flow which is maintained constant rather than the intraocular pressure (Bárány, 1947). The inadequacies of a purely mechanical theory mentioned earlier could be circumvented or remedied by the introduction of the neural control system. However, this addition to the theory carries the disadvantage that it must possess generality and explain the complex response of nerve sensors to adrenergic drugs, such as epinephrine.

Gaasterland et al. (1973) have investigated the effects of single doses of ℓ-norepinephrine and d,ℓ-isoproterenol in young men. Aqueous flow is reduced by beta-adrenergic stimulation combined with reduction of the pseudofacility from alpha-adrenergic stimulation. In particular, norepinephrine reduces pseudofacility without altering the intraocular pressure, episcleral venous pressure, or the true outflow facility. In contrast, isoproterenol reduces the intraocular pressure and aqueous outflow without changing the outflow facility or pseudofacility. Such drugs have an obvious application to clinical diagnosis of pathological eyes.

It is also known that sympathetic vasomotor tone can regulate the intraocular blood flow, which is linked in turn to the intraocular pressure. For example, using a nuclide-labelled microsphere technique for blood flow detection in cat eyes, Weiter et al. (1973) found that sympathectomy (extirpation of a 1 cm segment of the cervical sympathetic nerve) at constant intraocular pressure can increase ocular blood flow by more than 30%, while sympathetic stimulation leads to a decrease in flow rate of up to 50%. However, the experimental results do not pinpoint the sites in the arterial system at which sympathetic vasomotor control is exerted.

It is apparent that a complete theory of ocular dynamics in man must employ both passive (mechanical) and active (neural) elements to describe the variations of the intraocular pressure. Without a convincing means of separating the two and with the difficulties of characterising even the purely mechanical aspects described in previous sections, no attempt is made to do so in this preliminary study.

CHAPTER 5. MEASUREMENT TECHNIQUES

Several clinical measurement techniques are used on the human eye, each of which measures, directly or indirectly, the pressure in the eye. From these pressure measurements, values for ocular rigidity, ophthalmic blood pressure, and aqueous flow can be inferred. The clinical operation of these instruments will not be discussed, but a theoretical description of their operation will be presented in this chapter.

1. Measurement of intraocular pressure

A. Tonometry

If a finger is pushed into a balloon, the depth of penetration will depend upon the size and elasticity of the balloon, the force behind the finger, and the original pressure in the balloon, this pressure being elevated by penetration of the finger. Analogous behaviour holds in the eye. Indeed, the undisturbed intraocular pressure may be estimated by pressing a light plunger against the cornea or sclera.

The *tonometer* is a noninvasive instrument which estimates the ocular rigidty and the intraocular pressure by measuring the deformation produced by a known force or the force required to produce a known deformation. The compression of the cornea or sclera by a tonometer is equivalent to an injection of a small amount of fluid into the eye, resulting in an immediate rise in the intraocular pressure.

The three main types of tonometers encountered in clinical practice are shown in Figures 5.1a-c.

The *indentation* tonometer (Figure 5.1a) is a displacement-measuring device, with a constant force or weight applied on a known area. The most common indentation tonometer, indeed the most common tonometer of any type, is the Schiøtz tonometer, which typically has a plunger 1.5 mm in diameter. This plunger is balanced on the cornea, while the deformation of the cornea is recorded as a function of time. Four standard masses (5.5, 7.5, 10.0, and 15.0 gm) are used interchangeably with the Schiøtz tonometer to produce the constant force.

Figure 5.1a. Indentation (Schiøtz tonometer.

The *applanation* tonometer (figure 5.1b) is a force-measuring device which records the force necessary to flatten a specific area in the central cornea. The Goldmann applanation tonometer measures the force required to flatten a 3.06 mm diameter region. This diameter has been chosen so that the intrinsic resistance of the cornea against the tonometer is exactly balanced by the force of attraction created by the surface tension of the tears. The MacKay-Marg applanation tonometer flattens a small portion of the cornea but records the force operating only on a plunger situated concentrically within the overall applaning surface (Schwartz, 1966). In this case, the resistive force of the cornea is compensated by the annulus of the tonometer; the surface tension of the tears acts only around the periphery of this annulus and not upon the sensor in the centre.

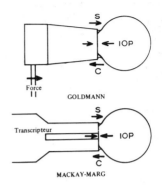

Figure 5.1b. Applanation (Goldmann and MacKay-Marg) tonometers.

The *suction cup* tonometer (Figure 5.1c) comprises a conical plastic cup which is placed over the entire front of the eyeball, the edge of the cup lying on the limbus. A syringe is used to create a low pressure within the cup so as to perturb the intraocular pressure without a local indentation of the cornea. This method does not permit observation of the eye during the pressure measurement.

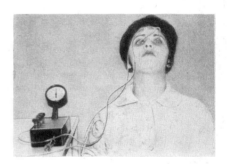

Figure 5.1c. Suction cup tonometer.

The intraocular pressure during *tonometry* depends upon two primary factors: the pressure in the eye before the tonometer is applied; and the ocular rigidity, i.e. the ease with which the corneo-scleral membrane stretches. In order to determine the ocular rigidity, it is necessary to make the measurements under paired conditions: with two weights using an indentation tonometer; with two areas using an applanation tonometer; or with a pair of readings using one applanation and one indentation tonometer.

It is important to note that a tonometer does not directly measure the steady state intraocular pressure but only the instantaneous pressure while it is balanced on the eye.

B. Tonography

Tonography is a longer duration version of tonometry, usually lasting four

minutes. Because of the raised intraocular pressure, the aqueous humour production
will be diminished and the outflow increased, both reactions tending to return the
pressure monotonically to its original equilibrium value. By suitable calibration,
the tonographic force or deformation measurement can be converted into the intraocular pressure. The resulting pressure curve, called a *tonogram* reflects the
combined effects of four factors: the cardiac pulse; the respiratory rate; Traube-
Hering variations in systemic blood pressure; and the monotonic pressure decay itself. In Section 3.2 we considered pulsations about the mean intraocular pressure.
Here, we concentrate on the exponential-like decay of the mean, as shown in Figure
5.2. Since the ocular rigidity can be inferred from short duration tonometry (of
the order of several seconds), tonography is used primarily to determine the facility
of aqueous outflow. The instrument used is invariably an electronic version of the
Schiøtz tonometer.

With known values of the ocular parameters, the model of Chapter 3 may be used
to predict a theoretical tonogram. Conversely, the same model may be used to eval-
uate unknown ocular parameters for a particular eye. Here, one would first estimate
the parameters, compute the theoretical tonogram, and readjust the parameters iter-
atively until the computed and clinical tonograms agreed satisfactorily. Special
precautions are necessary to guarantee uniqueness. The last term in Equation (3.1)
for external volume changes may be specified according to the type of tonometer
employed (Gloster, 1966).

Several basic assumptions are made in order to interpret the results from
tonography:

    (1) the rate of aqueous production remains constant;

    (2) the facility of aqueous outflow remains constant;

    (3) the ocular rigidity remains constant; and

    (4) prior to tonography, the eye is in a steady state.

In light of the literature survey in Chapter 2, only the last of these assumptions
is generally accurate.

During the early days of tonography, additional assumptions included constant
episcleral venous pressure and constant uveal blood volume. Over the years, however,
these restrictions were gradually dropped and the tonographic equation generalised
accordingly.

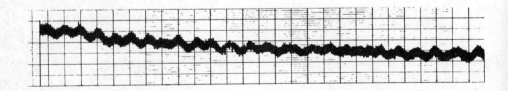

Figure 5.2. Sample tonogram. From Etienne (1965) - used by permission.

A comprehensive formula for the facility of aqueous outflow, which depends upon all four assumptions was given by Becker and Drews (1967, p. 50):

$$C_f = \frac{\Delta V_c + \Delta V_s + \Delta V_a + S_o t}{t(P_{av} - P_{vav})} ,$$ (5.1)

where ΔV_c is the indentation of the cornea, ΔV_s the distension of the sclera, ΔV_a the change in uveal blood volume, t time, P_{av} the average intraocular pressure during tonography, and P_{vav} the average episcleral venous pressure.

For clinical applications, a simplified version of this equation is generally used (Schimek, 1964, p. 16):

$$C_f = \frac{\Delta V}{t(P_{av} - P_o - \Delta P_v)} ,$$ (5.2)

where ΔP_v is the average increase in episcleral venous pressure during tonography, P_o the initial undisturbed intraocular pressure and

$$\Delta V = \Delta V_c + \Delta V_s .$$ (5.3)

ΔP_v is frequently taken to be 1.25 mmHg (Armaly, 1964). Note that the change in blood volume, ΔV_a, has now been assumed to be zero. Tables exist which relate the corneal indentation, ΔV_c, to the tonometer reading (Schimek, 1964). The scleral volume change, ΔV_s, is generally calculated clinically using Friedenwald's formula. Thus,

$$\Delta V = \Delta V_{c_2} - \Delta V_{c_1} + \frac{1}{MK} \ln (P_2/P_1) .$$ (5.4)

Equation (5.2) measures the increment in outflow per minute divided by the incremental pressure difference during tonography. Under assumption 2, this equals the normal facility of aqueous outflow.

For the corneal indentation, Shepherd et al. (1967) quoted the following from McBain (1960):

$$\Delta V_c = 79.75 - 2.06\ P$$ (5.5)

for a 5.5 gm mass, and

$$\Delta V_c = 64.84 - 1.293\ P$$ (5.6)

for a 7.5 gm mass, where P is pressure with the tonometer on the eye.

Armaly (1964) determined the effects of long duration tonography on human eyes. For most of his tonograms, the arithmetic average of the one-minute C_f value for each of the first four minutes did not differ from the four-minute value by more than 6%. He concluded that factors such as the reduction in intraocular blood volume, suppression of inflow, stress relaxation, and compression of the cornea by th tonometer plunger do not significantly affect the tonographic tracing. Levene and Hyman (1969), on the other hand, found that the facility was not constant over the entire tonographic period.

Processing of clinical Schiøtz tonograms is straightforward. The usual practice is to join the first and last points of the tonogram, it being understood that the curve of decreasing tonometer readings in time is approximated by the straight line joining these two points ("conventional" method). McEwen (1973) looked into the possibility of a better fit by either (a) using a chord through the first point which also achieves the smallest root mean squared deviation through the remaining points ("statistical" method) or (b) replacing the chord by a simple curve ("visco-elastic" method). Although small numerical differences result in values of the outflow facility calculated by the conventional and statistical methods, a better fit is obtained with the approximation of the tonogram by a curve.

Three other techniques allow an independent estimation to be made of the facility of outflow in a given eye: fluorescein injection into the antecubital vein; the rate of turnover of substances injected into the aqueous humour; and perfusion of the anterior chamber. In each technique, values for the facility of aqueous outflow are in close agreement with those determined by tonography (Becker and Drews, 1967, p. 58). These three techniques are unfortunately invasive.

Although tonography can demonstrate the presence of an abnormal resistance to aqueous outflow, it does not give information as to the site or nature of the obstruction (Becker and Drews, 1967, p. 37).

Spurious interpretations, particularly for abnormal eyes, may easily arise as a result of improper calibrations based on over-simplified theories. In addition, operator skill is necessary in effecting the measurement, as accidental decentering of the probe by as little as 1.5 mm can cause the loss of the characteristic tonographic form and raise the intraocular pressure by 4-5 mmHg. Smaller errors may be incurred if the probe is held at an improper angle with respect to the eye. Since the rigidities of the cornea and sclera are neither identical nor uniform, the calibration of the instrument also depends upon the precise region of application. In fact, in a large number of human subjects, scleral and corneal readings do not agree. Finally, if anaesthetics are to be avoided, the operator must select a relaxed instant so that the measured pressure is not artifically raised by the ocular muscles.

For indentation tonography, Equations (5.5) and (5.6) give the relation between $\Delta V_{ext} = \Delta V_c$ and $\bar{P}$ which we write in general form:

$$\Delta V_{ext} = d_1 - d_2 \bar{P}, \tag{5.7}$$

the derivative being

$$\frac{dV_{ext}}{d\bar{P}} = -d_2 . \tag{5.8}$$

The initial rise in intraocular pressure when the tonometer is placed on the eye is governed by Equations (2.46) and (5.7):

$$\ln \frac{\bar{P}_1 + b/a}{P_0 + b/a} = a\Delta V_{ext} = a(d_1 - d_2 \bar{P}_1), \tag{5.9}$$

where we again use McEwen and St. Helen's unifying form of the ocular rigidity function. This transcendental equation for the initial elevated pressure $\bar{P}_1$ can be solved by numerical iteration.

By combining Equation (5.8) with (3.12) and (3.16), the governing equation for indentation tonography is obtained:

$$\frac{d\bar{P}}{dt} = \frac{C_p(P_c - \bar{P}) - \dfrac{(1-a_3)\bar{P} - a_4}{a_1\bar{P} + a_2}}{\dfrac{1}{(a\bar{P} + b)(K_{eq}^* + K_f^* e^{-m_f t} + K_s^* e^{-m_s t})} + \dfrac{V_a}{K_\alpha (P_a - \bar{P})^\alpha} + d_2}, \tag{5.10}$$

with the pressure at time t = 0 being $\bar{P} = \bar{P}_1$ as determined from Equation (5.9). Equation (5.10) must be integrated numerically. Figure 5.3 shows the calculated results of the tonographic behaviour for our Standard Eye with both 5.5 and 7.5 gm plunger weights.

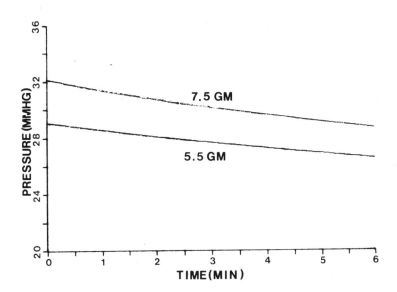

Figure 5.3. Model results for Schiøtz indentation tonography. Upper curve is for 7.5 gm mass; lower is for 5.5 gm mass.

C. Perilimbal suction cup method**

The perilimbal (or scleral) suction cup method blocks the outflow of aqueous so that the intraocular pressure rises due to continued aqueous secretion. The suction cup, with 50 mmHg negative pressure, is applied to the limbal area of the eye (Figure 5.1), compressing the episcleral and intrascleral vessels. There is an immediate *rise* in intraocular pressure due to distortion of the eye by the suction cup. McBain *et al.* (1967) correlated this pressure rise with the undisturbed intra-ocular pressure.

$$\Delta P = 17.0 - .25 \, P_o. \tag{5.11}$$

Adding P_o to each side of equation (5.11) gives

$$P_{SC_1} = 17.0 + .75 \, P_o, \tag{5.12}$$

where P_{SC_1} is the "initial" intraocular pressure with the cup applied.

After the initial rise in pressure and while the cup is still on the eye, two events occur: the sclera relaxes because of its viscoelasticity, tending to reduce the pressure; and continued aqueous secretion acts to increase the pressure. Of the two influences, the latter is stronger, so that there is a continual net rise in intraocular pressure as long as the cup remains on the eye, typically 10 minutes.

Removal of the cup reopens the outflow channels, and the pressure gradually falls back to its normal value in about 20 minutes. An immediate drop in pressure occurs when the cup is removed because the eye returns to its customary undistorted shape. We assume that an equation similar to Equation (5.11) also governs this case. By rearranging Equation (5.11), we can calculate the pressure drop $|\Delta P_f|$ at the end of the 10-minute period with the cup still on the eye from P_{SC_2}, the known pressure at that time:

$$|\Delta P_f| = 22.7 - .33 \, P_{SC_2} . \tag{5.13}$$

Subtraction of $|\Delta P_f|$ from the final intraocular pressure (P_{SC_2}) just prior to re-moval gives an "initial" elevated pressure analogous to the initial pressure in tonography:

$$P_1 = P_{SC_2} - |\Delta P_f| = 1.33 \, P_{SC_2} - 22.7. \tag{5.14}$$

An important difference in this case, of course, is that no external force acts on the eye as in tonography. McBain *et al.* (1967) gave an equation for the pressure immediately after cup removal which, unfortunately, fits none of their plotted data. Since they did not show how they derived that equation, it is impossible to trace the source of this discrepancy.

Except during the virtually instantaneous application and removal of the suction cup, there is no change in the externally imposed volume during suction

---

** This section is derived principally from Schimek (1964) and McBain *et al.* (1967).

cup tonography, i.e. $dV_{ext}/dt = 0$. Also, since the aqueous outflow is suppressed completely, Equation (3.16) takes the simpler form:

$$\frac{d\bar{P}}{dt} = \frac{C_p(P_c - \bar{P})}{\dfrac{1}{(a\bar{P} + b)(K_{eq}^* + K_f^* e^{-m_f t} + K_s^* e^{-m_s t})} + \dfrac{V_a}{K_\alpha(P_a - \bar{P})^\alpha}} . \qquad (5.15)$$

The initial intraocular pressure with the suction cup applied to our Standard Eye is obtained from Equation (5.11) and equals 28.625 mmHg. Computed results are shown in Figure 5.4. The pressure drop at t = 10 min when the suction cup is removed is calculated from Equation (5.14).

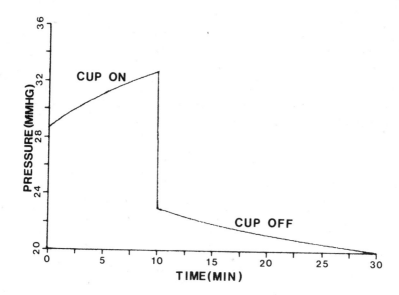

Figure 5.4. Model results for suction cup tonography. Note the different time scale from Figure 5.3.

Recognition that disorder of the carotid arteries can cause transient cerebral ischaemia and strokes and that these disorders can be prevented by surgical inter- vention has enhanced the value of diagnostic procedures which assess the adequacy of the blood supply to the brain.   Direct flow measurements in the cerebral circulation have traditionally been both difficult and hazardous, although recent advances in transcutaneous Doppler flowmeters are rapidly reversing this situation.   Pressure measurements have generally been more feasible; namely, in the eye and the arm. (Figure 6.1)

Knowledge of the blood pressure in the ophthalmic artery contributes to the clinical investigation of cerebral vascular disturbances, glaucoma, and arterial hypertension.   Since ophthalmic arterial pressure depends on the systemic arterial pressure and follows its general variations, the ophthalmic/brachial arterial pressure ratio (OAP/BAP) is more significant than the ophthalmic arterial pressure alone.   This ratio, both diastolic and systolic, varies with pressure levels and age.   A few examples are cited here from the work of different researchers: Bailliart (1917) - diastolic OAP/BAP = 45%; Streiff (1937) - diastolic 50%, systolic 70%; Perry and Rose (1958) - diastolic 73% in normotensives, 81% in hypertensives; Borrás *et al.* (1969) - diastolic 92.8%, systolic 75.6%, these last results being direct pressure measurements by catheterisation.

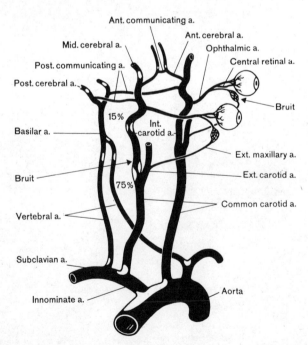

Figure 6.1.   Upper systemic arteries showing common sites of occlusion.   From Hager (1964) - used by permission.

Weigelin and Lobstein (1963) related the mean values of ophthalmic and brachial arterial pressures to the intraocular pressure:

$$\bar{P}_{oph} = \gamma \, (P_{brach} - P) + P. \qquad\qquad (6.1)$$

The constant γ is reported on different occasions as 0.475 and 0.73, although the second value seems to be more consistent with normal physiological ophthalmic arterial pressures.

The brachial arterial pressure provides a baseline, related to the carotid pressure, to which the ophthalmic arterial pressure may be compared. In hypotensive patients, i.e. those with a low systemic arterial pressure, one would expect a corresponding decrease in ophthalmic arterial pressure; hence, the importance of the *ratio* OAP/BAP.

1. Clinical methods of assessing the cerebral circulation

Any nontraumatic technique which provides information about the circulation of the brain is valuable. This section will describe several classical and newer techniques for assessment of the cerebral circulation.

A. Ophthalmodynamometry

In an extension of the tonometry procedure, the tonometer probe is pressed harder against the eye, causing a greater elevation of the intraocular pressure, which in turn constricts the central retinal artery, causing it to pulsate markedly and, subsequently, to collapse as the intraocular pressure exceeds the arterial pressure. The initiation of marked pulsations, directly observable through an ophthalmoscope, indicates the diastolic pressure, while cessation of all pulsations and complete collapse of the artery indicate the systolic pressure (Bailliart, 1917; Koch, 1945).

Ophthalmodynamometry provides an indication of the systolic and diastolic pressures which existed in the central retinal artery before the procedure was performed. However, the blood flow is impeded or stopped during this measurement, and the pressures obtained pertain not to the central retinal artery itself, but rather to some more distant point in the larger ophthalmic artery. Van der Werff (1972) used a steady Poiseuille flow analysis to provide a preliminary estimate of location. He estimated that the normal retinal arterial systolic pressure was 14 to 17 mmHg lower than the systolic pressure determined by ophthalmodynamometry, this discrepancy arising due to the large pressure gradient caused by the small lumen of the central retinal artery when blood flows through it normally. Bill (1963) had shown this experimentally in rabbits, for which he found a pressure difference of 14 mmHg at the origin of the ciliary artery.

The significance of ophthalmodynamometry in screening for cerebrovascular disease is that differences greater than 15% in the central retinal pressures in the two eyes indicate a compromised blood flow due to partial occlusion of the internal carotid artery on the side of the lower pressure (Perry and Rose, 1958).

Hayreh (1963) notes that obstruction of the internal carotid artery below the origin of the ophthalmic artery (Figure 6.1) leads to a fall in pressure whereas an obstruction above the origin leads to a rise in pressure. He concludes:

> Ophthalmodynamometry has a great future, being a simple, harmless, reliable, readily available, and rapid test in diagnosing arterial occlusion at various levels between the aortic and the terminal branches of the internal carotid artery, including the ophthalmic artery and the central retinal artery, and in evaluating the condition of the cerebral circulation, as well as its equally important role in the treatment and prognosis of cerebrovascular occlusive disorders.

B. Ophthalmodynamography

Hager (1964) suggested a device for measuring the ophthalmic arterial pressure which functions on the same principle as Bailliart's ophthalmodynamometer, with the following important differences: (1) a pneumatic cup placed over the eye applies positive pressure, in contrast to Bailliart's mechanical plunger pressed against the eye; and (2) instead of visual observation of the pulsations of the central retinal artery (which is no longer possible, unless a window is fitted into the cup), the pressure variations in the cup are recorded electronically. Hager described the technique as follows:

> In the same way as with the sphygmomanometer, a hand-bellows is used to pump an ophthalmic pressure chamber and a shunted sphygmomanometer cuff to a pressure exceeding the systolic pressure. A pressure nozzle, built into the ophthalmodynamograph, allows the pressure to be gradually reduced at a regular rate. During this process, an automatic pressure gauge generates impulses at intervals of 20 or 10 mmHg, which impulses are registered in the form of a cardiograph together with the oscillogram trace of the ophthalmic artery. At the same time, a microphone, fixed at the crook of the elbow, registers the Korotkoff sounds of the brachial artery. In this way synchronous, automatic determination of the blood pressure in the upper arm and the blood pressure of the ophthalmic artery is made possible. The pulsation volume of the ophthalmic artery is determined by comparison of maximum oscillations and a known control volume. Moreover, if the carotid pulse is simultaneously recorded, the time-lag of the "ophthalmic pulses" enables the pulse wave velocity of the internal carotid to be obtained. This latter, in turn, provides us with further information on the degree of elasticity of the cerebral vessels.

Since the cup obscures direct viewing of the pulsations of the central retinal artery during ophthalmodynamography, Hager proposed the following criteria for

67

defining systole and diastole (Figure 6.2):

The schematic representation of the oscillogram trace enables the contour
criteria of systolic and diastolic pressure in the ophthalmic artery to be
clearly seen. Where the cushion pressure of the ophthalmic pulse is in excess
of the blood pressure we see nothing more than small residual deflections
produced by the impact and reflection of the pulse waves at the occluded
section of the vessel. When the cushion pressure sinks below the systolic
blood pressure, or reaches it, the vessel opens during the systolic pressure
peak very slightly and for a very short time; the natural reflection of this
in the oscillogram is at first a higher peak oscillation. Once the systolic
pressure peak is past, the cushion pressure is in the ascendant and forces the
vessel to collapse until the next systolic pressure peak occurs. The vascular
collapse, which is progressively curtailed as the cushion pressure sinks, is at
first reflected in the oscillogram as a progressively-diminishing flat at the
base of the cycle. Finally, there comes a point at which the cushion pressure
no longer collapses the vessel completely. In this range, the vascular wall
itself is now almost completely unstressed and can therefore follow the
fluctuations of the pulsatory pressure volume practically without inhibition.
This explains the fact that in the diastolic range, maximum oscillations
appear and the descending limb of the cycle passes instantaneously into the

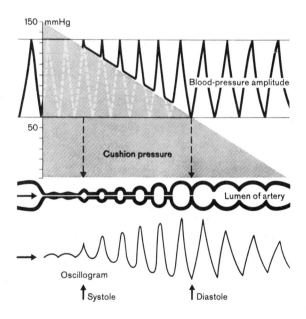

Figure 6.2. Determination of systolic and diastolic pressures in Hager ophthalmodynamography. From Hager (1964) - used by permission.

ascending limb of the following cycle, i.e. the flat at the base of the cycle
now becomes an acute apex or true cyclic foot. Thus the diastolic value is
attained when the previously flat or rounded bases of the oscillations sharpen
to points. At values below the diastolic value these maximum oscillations
become appreciably smaller; this may take place immediately or after the next
two or three pulsations. Similar criteria obtain in respect of blood pressure
in the upper arm. The Korotkoff sounds, corresponding to the systolic pressure,
are reflected in either a sudden reduction in amplitude or the wave or in its
total disappearance.

C. Ocular pulse analysis

Galin *et al.* (1972) suggested that the ocular pressure pulse might be useful
in the detection of obstructive lesions in the carotid tree. The ocular pulse can
be recorded by a pressure transducer connected via fluid-filled polyethylene tubing
to a suction cup applied to the eye globe at the limbus. The amplitude and the area
of the pulse wave, its onset relative to the R-wave of the electrocardiogram, slopes
of the pressure ascent and descent legs, and the dicrotic notch position can be
correlated with the results of such classical clinical techniques as ophthalmodynamometry and angiography. The above factors are altered in the presence of carotid
occlusion or stenosis and can form the basis of a diagnostic detection system.

Galin *et al.* (1972) claimed an advantage for ocular pulse analysis relative to
ophthalmodynamometry in diagnosis of stenosed carotids. For animals with induced
progressive lesions of the carotid tree, they found that detectable changes in the
ocular pulse occurred at 50% reductions in lumen cross-section, whereas they
required an 85% reduction in cross-section for the lesion to be detected by ophthalmodynamometry.

Ocular pulse analysis has been dramatically enhanced by Best and Rogers (1974)
by submitting the ocular pulse to Fourier analysis to give

$$P = M_0 + \sum_{n=1}^{N} M_n \sin(n\omega t + \phi_n), \tag{6.2}$$

where M_n and ϕ_n are the magnitude and phase of the nth harmonic, respectively, ω the
angular frequency, and N the number of harmonics calculated. Usually, $N = 3$.

Best and Rogers used this technique to examine occlusions of the common carotid
in rabbits, the stenoses being produced with a microcaliper. They found that the
magnitude of the first harmonic was reduced significantly in all degrees of stenosis,
the second harmonic was reduced significantly in 50% or greater degrees of stenosis,
and that the third harmonic was affected unpredictably in all degrees of stenosis.
Using the standard graphical techniques of ophthalmodynamography, stenoses of only
50% or greater degree could be detected. With Fourier analysis, however, stenoses
as small as 20% could be diagnosed.

D. Carotid compression

Tonographic recordings of intraocular pressure variations during temporary digital compression of the common carotid artery on the lower neck proximal to the carotid sinus constitute another valuable technique for the detection of carotid occlusions. It is evident from Figure 6.1 and from the discussions in Chapter 2 that a constriction of the internal carotid artery will lead to a decrease in the blood flow to the eye and in the ophthalmic and intraocular pressures. Thus, simultaneous measurements of intraocular and carotid arterial pressures provide a diagnostic test for carotid occlusion.

The normal response to compression of the common carotid for one or two seconds is an abrupt drop in intraocular pressure on the compressed (homolateral) side, with a negligible drop on the opposite (contralateral) side. The intraocular pressure returns to its original value almost immediately upon releasing the carotid (Galin *et al.*, 1969).

In the presence of carotid occlusion, the intraocular pressure falls on the occluded side upon compression of the normal common carotid, with only a minimal drop on the healthy side. When the occluded carotid is compressed, the intraocular pressure remains almost invariant (Neuschüler and Pepe, 1967).

This indicates a collateralisation from the normal to the abnormal side as indicated in Figure 6.3. When carotid A is compressed, the pressure falls in B, indicating the presence of an occlusion and collateralisation; whereas compression of B will not cause a large pressure drop in B, since it is supplied by A which continues unhindered. The presence of contralateral intraocular pressure fall on

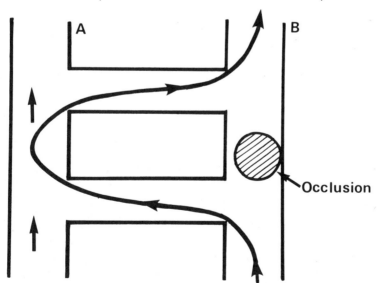

Figure 6.3. Schematic indication of collateralisation during carotid occlusion.

homolateral common carotid compression has proven to be the most clinically
significant finding in the common carotid compression test (Neuschüler and Pepe,
1967).

When both carotids are occluded, the test indicates from which side the
collateralisation is greater. For example, left pressure fall on right compression
indicates that collateralisation is proceeding from right to left. Surgical inter-
vention on the right side, which is supplying a larger portion of the blood flow,
should then be avoided. Unless the compression test is performed proximal to the
carotid sinus, bilateral changes may occur which will mask the diagnosis.

The compression technique also enables one to differentiate between occlusion
and stenosis of a carotid by using ophthalmodynamography (cf. Section B above). If
the central retinal arterial pressure of the affected side (already suspect) drops
after homolateral compression, a stenosis is indicated; if the central retinal
arterial pressure drops only after contralateral compression, an occlusion is
present.

The carotid compression procedure is not without risk. Silverstein *et al.*
(1959) tested forty patients with occlusive disease, the compression being maintained for 30 seconds. The majority of patients experienced both seizures and
syncope when their patent carotids were compressed. In isolated cases, carotid
compression may dislodge atheromatous plaques into the circulation. These latter
complications are fortunately rare, and it is generally felt that the information
to be gained from careful performance of the procedure outweighs the risks.

The use of an electroencephalogram (EEG) during the procedure further increases
its safety. In particular, the EEG, which measures the electrical activity of
specific regions of the brain, may indicate abnormal patterns if carotid compression
begins to compromise an adequate supply of blood to the brain, in which case the
test may be abruptly discontinued.

E. Other methods

Additional diagnostic procedures for assessing cerebrovascular pathology will
be described only briefly.

Thermography detects surface temperature differences by infrared photography.
It is used extensively for the detection of breast cancer, in which the tumor's
higher metabolic rate produces a local rise in temperature. Wood *et al.* (1966)
applied the technique to diagnosis of cerebrovascular occlusive disease. Heat is
carried to the forehead by blood flow in the frontal and suprafrontal arteries.
Temperature differences on the two sides of the forehead of less than 0.2°F are
normal, whereas differences exceeding 0.6°F indicate a compromised blood flow on
the side of the lower temperature. Certain forms of dermatitis may also alter the
skin temperature, complicating the diagnosis, but the noninvasive nature of the
technique renders it of great potential value.

Doppler sonography is another noninvasive method for assessing carotid artery disease. High frequency sound waves are directed at an artery and reflected by the moving red blood cells and by the pulsating arterial wall. As checked against angiography, Hyman (1974) correctly determined the patency or obstruction of the internal carotid artery in 83% of patients (N = 53) using a Doppler flowmeter. All but one of the mismatches were false positives.

Angiography, involving X-ray observation of radio-opaque dye injected into the bloodstream, still constitutes the most reliable method of detecting anatomical deficiencies in the cerebral circulation in spite of the hazards involved. We suggest that the more innocuous techniques should be used first, with angiography being reserved for the most serious cases.

CHAPTER 7. SUMMARY AND CONCLUSIONS

In this monograph the important fluid mechanical and viscoelastic aspects of ocular dynamics have been presented, with the main interest centering on the regulation of intraocular pressure. Aqueous humour formation and outflow rates largely determine the state of equilibrium achieved. Neural control, not extensively covered here, probably acts through vasomotor adjustment of the ocular blood flow, which is coupled to both aqueous formation and outflow.

Present knowledge of the eye's mechanical characteristics has been gathered from a comprehensive literature survey and the empirical findings translated into mathematical form. These relationships have been depicted in an "influence diagram".

Several models for the ocular rigidity function have been derived or discussed. The linear elastic model gives a linear pressure-volume relation for the eye, but is valid only for small changes from the steady state pressure. Friedenwald's logarithmic model is the one most often used clinically, but it too becomes invalid for large changes from the steady state. McEwen and St. Helen's unifying formulation of ocular rigidity is valid over the complete physiological range of intra-ocular pressures and was ultimately included in our mathematical model.

This model accounts for changes in aqueous production and outflow rates due to intraocular pressure changes, stress relaxation in the corneo-scleral wall, and forces externally imposed during measurement procedures. A "coefficient of aqueous production", analogous to the common coefficient of aqueous outflow, has been introduced to explain the dependence of aqueous production upon the pressure.

A nonlinear differential equation governing the time rate of change of the intraocular pressure has been formulated, employing a judicious selection of the sometimes inconsistent and incomplete experimental data available in the literature. Numerical integration of this equation provides a means of assessing the influence of various properties, such as ocular rigidity and outflow facility, on the intra-ocular pressure. We emphasise that such results as have been presented are preliminary and exploratory in nature.

Clinical methods for measuring pressures in the eye have been reviewed briefly and the important connection between ocular dynamics and blood flow in the brain described. We suggest that exploitation of known results and vigorous pursuit of new information on the dynamic behaviour of the eye can provide an invaluable "window" into the cerebral circulation.

We hope that the present monograph may contribute both to an improvement in accuracy and reliability of current clinical methods for diagnosing disorders of the eye and to an advancement of our basic understanding of ocular mechanics. In particular, the mathematical model can provide a valuable framework for the design of experimental and clinical tests and the interpretation of their results.

GLOSSARY

This glossary is provided for readers who may find some of the vocabulary in this monograph unfamiliar. In many cases the definitions have been specialised to the eye, even though they may also pertain to other structures in the body. For example, "limbus" is a general anatomical term denoting the border between adjacent structures, but we define it here for the specific border in the eye associated with that name.

accommodation: adjustment of the lens of the eye in order to focus at different distances

amaurosis fugax: momentary loss of vision due to spasm of the central retinal artery or partial occlusion of the internal carotid artery

anastomosis: connection between two blood vessels

angiography: X-ray visualisation of blood vessels

anterior: in front of

anterior chamber: the aqueous-filled space between the cornea and the iris

applanation: local flattening; for example, of the cornea during tonography

aqueous humour: the clear, watery fluid produced in the eye, occupying the anterior and posterior chambers

arteriole: a minute arterial vessel just proximal to a capillary

artery: a vessel through which the blood flows away from the heart to the various parts of the body

autoregulation: the property of a process which returns it automatically to its initial state after having been subjected to perturbation

axon: the central, conducting part of a nerve; many axons comprise the nerve fibre layer of the retina and converge to form the optic nerve

brachial artery: distributes blood to the shoulder, arm, forearm, and hand

cerebrovascular: pertaining to blood vessels of the brain

choroid: the thin, dark brown, vascular lining of the posterior five-sixths of the eyeball

ciliary: used particularly in reference to certain structures in the eye, as the ciliary muscle, ciliary process, and ciliary ring

circle of Hovius: an intrascleral circular arrangement of anastomosing ciliary veins near the limbus; not found in man

coefficient of ocular rigidity: ratio of intraocular pressure change to intraocular volume change, i.e. dP/dV; commonly expressed in the units $mmHg/\mu l$

coefficient of vascular rigidity: the ratio of pressure change to volume change in a blood vessel or vascular bed, i.e. dP_a/dV_a

collagen: a stiff protein providing the main structural support to skin, tendon, bone, cartilage, and connective tissue; can be converted into gelatin by - boiling

cornea: the transparent structure at the front of the eye, consisting of five distinct tissue layers

corneo-scleral envelope: the complete outer covering of the eyeball, comprising the cornea and sclera creep: the gradual distension of a body subjected to a constant stress

deformation: the change of form, shape, or especially the size of a structure

dialysis: the process of separating substances or ions in solution by the difference in their rates of diffusion through a semi-permeable membrane

diastole: the period of relaxation of the heart between contractions, during which the heart is refilled with blood from the veins

diffusion: the process of mixing of liquids or gases as a result of local gradients in their concentrations

distal: downstream

Doppler: the phenomenon by which the pitch (frequency) of a whistle (emitter) on a moving body (like a locomotive or a red blood cell) is higher when the body approaches the listener (receiver) and is lower when the body moves away; ultrasonic blood flowmeters are based on this principle

elastin: a flexible connective protein, considerably more elastic than collagen

empirical: pertaining to a relationship describing experimental results, without reference to underlying scientific principles

endothelium: the inner lining of the heart, vessels, and small cavities

enucleation: removal of the whole eyeball after the eye muscles and optic nerve have been severed

episcleral: overlying the sclera

epithelium: the outer lining of the heart, vessels, and small cavities

facility: an ocular parameter characterising the ease of flow (reciprocal of resistance); defined as the ratio of the flow rate to the pressure difference driving it; for example, outflow facility of aqueous humour from the anterior chamber of the eye

footplate: the flat or disc-like portion of a tonometer in contact with the eye

Friedenwald's coefficient: the constant of proportionality between the intraocular pressure and the *slope* of the pressure curve

ganglion cells: contained in the innermost of the nine layers of the retina; these cells pass preprocessed visual information to the brain

glaucoma: includes a complex set of disease entities which result in defects in the visual field due to degeneration of the optic disc and/or nerve; the condition, due to insufficient blood supply to the nerve fibres, is usually accompanied by an elevation of the intraocular pressure; various forms include primary glaucoma, congenital glaucoma, secondary glaucoma, and absolute glaucoma

Goldmann tonometer: a tonometer based on the principle of applanation

homeostasis: the tendency of physiological systems to maintain a stable state in spite of disturbances or changes in environmental conditions

in situ: in the natural or normal place; confined to the site or origin without extraction or invasion of neighbouring tissues

intraocular: within the eye

in vitro: outside the natural environment of the body; literally, "within a glass"

in vivo: within the living body

iris: the circular pigmented membrane behind the cornea, perforated by the pupil

lamina cribrosa: a sieve-like opening (0.7 mm long, 1.5 mm in diameter) in the posterior sclera through which passes the optic nerve lens: the transparent biconvex body of the eye situated between the posterior chamber and the vitreous body, consituting part of the refracting mechanism of the eye

limbus: the junction of the cornea with the sclera

membrane: a thin layer of tissue covering a cell or other tissue

modulus of elasticity: the ratio of stress to strain

neural: pertaining to a nerve or nerves

occlusion: collapse or closure of a vessel, e.g. of an artery through which blood flow is blocked

ocular: pertaining to the eye (Latin: *oculus*: eye)

ophthalmic: pertaining to the eye (Greek: *ophthalmus*: eye)

ophthalmodynamometry: a procedure for determination of the blood pressure in the retinal artery

optic: pertaining to the eye (Greek: *optikos*: of or for sight)

optic disc: ophthalmoscopically visible portion of the optic nerve

optic nerve: a nerve trunk consisting of about one million axons arising from the ganglion cells of the retina

osmosis: the selective passage of solvent across a membrane which impedes the passage of solute molecules but is permeable to the solvent

perfusion pressure: the pressure drop as a fluid passes through the vessels of a specific organ or body part; the driving pressure

perilimbal: surrounding the limbus

phasic: periodic or rhythmic

plunger: cylindrical part of a tonometer

Poiseuille's law: the volume flow rate in a circular tube is proportional to the fourth power of its radius; rigorously valid only for fully-developed flow in a long, rigid vessel in which inertial and viscous florces are in equilibrium and for which flow profiles no longer vary with distance along the vessel; rarely realised in man, except perhaps occasionally in the microcirculation

Poisson's ratio: an elastic parameter which relates the lateral deformations of a solid following deformation in the normal (orthogonal) direction; for most bio-logical soft tissue, this ratio is nearly one-half, indicating that the total volume of the tissue does not change during deformation

posterior: behind; in back of

posterior chamber: the aqueous-containing space between the iris and the lens

presbyopia: loss of elasticity of lens, generally with advancing age, causing the near point of distinct vision to be removed farther from the eye

proximal: upstream

pseudofacility: that component of the total facility related strictly to the formation of aqueous humour, characterised as the rate of suppression of aqueous inflow divided by the corresponding change in intraocular pressure

pulse wave speed: the speed of propagation of pressure waves along the blood vessel

pupil: the opening in the centre of the iris which allows light into the eye

retina: the innermost layer of the eyeball, surrounding the vitreous body, and connected to the optic nerve; responsible for the visual behaviour of the eye

retinal circulation time: an index of retinal circulation, calculated as the difference between mean venous and arterial times of arrival of labelled substances in the blood

sclera: the tough white tissue covering approximately the posterior five-sixths of the eyeball

Schiøtz tonometer: a tonometer based on the principle of indentation

Schlemm's canal: a circular channel at the limbus for aqueous outflow

secretion: the process of producing a substance, typically liquid, in a biological system

Standard Eye: the reference eye used in this text possessing average human characteristics

steady state: a state characterised by a non-varying mean value, even though there may be cyclic variations about the mean

stenosis: local narrowing of a vessel

strain: a nondimensional form of deformation in which the latter has been divided by the dimension of the body prior to deformation

stress: force per unit area

stress relaxation: the gradual decay in stress in a body subjected to a constant strain

stroma: the underlying structural matrix of a tissue

suction cup: placed over the eye and pressurised in one form of ophthalmodyna-mometry or depressurised in suction cup tonometry

systole: the period of contraction of the heart, during which blood is pumped into the aorta and the pulmonary artery

tension: act of stretching, usually synonymous with stress or pressure, as in intraocular tension, but sometimes defined as force per unit length

thermography: detection of surface temperature of the body, based on self-emanating infrared radiation; sometimes used as a noninvasive means of detecting tumours

time constant: a characteristic time; often the time required for an exponentially varying relaxation process to change by 1/e (e = 2.718) of the difference between its initial and final values

tonogram: the record produced by tonography

tonography: the recording of changes in intraocular pressure following the application of a known pressure, weight, or deformation on the eye globe, reflecting primarily the facility of outflow of the aqueous humour from the anterior chamber

tonometer: an instrument for measuring tension, pressure, or deformation in the eye

trabecular network: a mesh-like structure through which the aqueous drains out of the anterior chamber and into the veins

transmural: pertaining to difference measured across the wall of a vessel or organ

Traube-Hering waves: rhythmic variations in the arterial pressure due to the activity in the vasomotor centre

ultrafiltration: filtration under pressure through filters or membranes with minute pores; used to separate a substance in colloid solution from its dispersion medium

uvea: the iris, ciliary body, and choroid considered together

vascular: pertaining to or full of blood vessels

vasoconstriction: the diminution of the calibre of vessels; especially constriction of arterioles leading to decreased blood flow through that vessel vasodilation: dilation of a vessel; especially dilation of arterioles leading to increased blood flow through that vessel

vasomotor: any element or agent that effects the calibre of a blood vessel

vein: a vessel through which blood passes from various organs or tissues back to the heart

viscoelasticity: denoting the property of a tissue or deformable solid to distend in a manner which depends not only upon the force exerted, but also upon the rate of application of this force; the gradual relaxation behaviour of biological soft tissues under constant stress or strain

vitreous: the transparent substance that fills the part of the eyeball between the lens and the retina

MATHEMATICAL NOTATION

In order to give a unique designation to each varibale used in this monograph, many of the equations found in the literature have been rewritten in a slightly different but equivalent form. The number in parentheses after each description refers to the page where the variable first appears in this monograph.

a	Constant (30)
a_1	Constant (46)
a_2	Constant (46)
a_3	Constant (46)
a_4	Constant (46)
A	Area of arterial lumen corresponding to radius r (15); filtration area (27); constant (36)
b	Constant (31)
c	Constant (31)
C_f	Facility of aqueous outflow (25)
c_o	Moens-Korteweg pulse wave speed (18)
C_p	Facility of aqueous production (21)
C_{tot}	Total facility (28)
d_1	Constant (60)
d_2	Constant (60)
E	Rigidity ($\equiv dP/dV$) (31)
E_a	Elastic modulus of arterial wall (15)
E_e	Elastic modulus of corneo-scleral wall (34)
E_{inc}	Incremental modulus of elasticity (17)
h	Wall thickness of artery (15)
H	Wall thickness of sclera (33)
I	Infusion rate (22)
k	Proportionality factor (17); matrix permeability (24)
K	Friedenwald's coefficient of ocular rigidity (30)
K_a	Proportionality factor (16)
K_{eq}	Equilibrium rigidity constant (38)
K_{eq}^*	Normalised equilibrium constant (40)
K_f	Fast rigidity constant (38)
K_f^*	Normalised fast rigidity constant (40)
K_s	Slow rigidity constant (38)
K_s^*	Normalised slow rigidity constant (40)
K_t	Total rigidity constant (40)
K_T	Rigidity constant at time T (48)
K_o	Proportionality factor (34)
K_α	Proportionality factor (17)
ℓ	Length (25)

L	Hydraulic conductivity (27); length (35)
m_f	Time constant (38)
m_s	Time constant (38)
M	Conversion factor between logarithms (7)
M_n	Magnitude of harmonic (68)
M_o	Mean of ocular pulse (68)
n	Exponent (31); summation index (68)
N	Number of harmonics (68)
P	Intraocular pressure (11)
$\bar{P}$	Mean intraocular pressure (48)
P*	Pulsating component of intraocular pressure (48)
P_a	Arterial pressure (15)
P_{av}	Average intraocular pressure during tonography (25)
P_{brach}	Brachial artery pressure (65)
P_c	Cutoff pressure (21)
P_{cap}	Capillary pressure (27)
P_{cH}	Circle of Hovius venous pressure (11)
P_{CO_2}	Partial pressure of carbon dioxide (9)
P_{fem}	Femoral artery pressure (13)
P_m	Pressure in matrix (24)
P_o	Mean steady state intraocular pressure (12)
P_{oph}	Ophthalmic artery pressure (14)
P_{osm}	Osmotic pressure difference between blood and aqueous (27)
P_{O_2}	Partial pressure of oxygen (9)
P_{ret}	Retinal artery pressure (44)
P_{sc}	Intraocular pressure during suction cup procedure (62)
P_{syst}	Systolic pressure determined during ophthalmodynamography (44)
P_v	Episcleral venous pressure (12)
P_{vav}	Average episcleral venous pressure during tonography (59)
$\vec{q}$	Volume flow rate of fluid per unit area (25)
Q	Blood flow rate (9); aqueous flow rate (25)
Q_{aq}	Aqueous flow rate (43)
Q_{in}	Aqueous humour inflow (27)
Q_{out}	Aqueous humour outflow (27)
Q_u	Uveal blood flow rate (13)
Q_1	Aqueous flow rate due to ultrafiltration (27)
Q_2	Aqueous flow rate due to active secretion (27)
Q_3	Aqueous flow rate due to percolation through trabecular network (27)
Q_4	Aqueous flow rate due to uveoscleral bulk flow (27)
r	Inner radius of artery (15)
R	Inner radius of eye (33)

RCT	Retinal circulation time (10)
R_f	Resistance to aqueous outflow (25)
R_{20}	Resistance to aqueous outflow at P = 20 mmHg (26)
S_o	Aqueous humour outflow rate (25)
S_p	Aqueous humour production rate (21)
t	Time (25)
T	Specific time (48)
v	Velocity (44)
V	Intraocular volume (29)
V_a	Vascular volume (16)
V_{aq}	Aqueous volume (43)
V_{ext}	Volume changes imposed upon the eye (45)
VR	Coefficient of vascular rigidity (18)
x	Dummy variable (7)
α	Arterial elasticity exponent (17)
β	Arterial elasticity exponent (17); corneo-scleral elasticity exponent (35)
γ	Proportionality factor (65)
ε_a	Strain in arterial wall (15)
ε_e	Strain in corneo-scleral wall (33)
μ	Fluid viscosity (24)
ν_a	Poisson's ratio for arterial wall (15)
ν_e	Poisson's ratio for corneo-scleral wall (34)
ρ	Blood density (18)
σ_a	Circumferential stress (15)
σ_{ch}	Choroidal stress (36)
σ_e	Stress in corneo-scleral wall (33)
σ_{scl}	Scleral stress (36)
ϕ_n	Phase of harmonic (68)
ω	Angular frequency of blood pulse (68)
ΔA	Area change (15)
ΔP	Intraocular pressure change (17)
ΔP_a	Vascular pressure change (17)
ΔP_{cc}	Critical closure pressure (16)
ΔP_{eq}	Equilibrium pressure difference (38)
ΔP_f	Final pressure fall with suction cup removal (62)
ΔP_{fast}	Fast relaxation component of pressure change (38)
ΔP_{slow}	Slow relaxation component of pressure change (38)
ΔP_v	Increase in episcleral venous pressure during tonography (25)
ΔR	Ocular radius change (33)
ΔV	Intraocular volume change (7)

ΔV_a Arterial bed volume change (16)

ΔV_c Corneal indentation (25)

ΔV_{ext} Externally imposed volume change (60)

ΔV_s Scleral distension (59)

Subscripts:

o Reference condition; steady state value; state at zero transmural pressure (15)

1 Initial state (17)

2 Final state (17)

REFERENCES

F.H. ADLER (1965). *Physiology of the eye*. 4th ed. C.V. Mosby, St. Louis.

B. ANDERSON Jr. & H.A. SALTZMAN (1964). Retinal oxygen utilization measured by hyperbaric blackout. Arch.Ophth. 72, 792-795.

D.R. ANDERSON & W.M. GRANT (1973). The influence of position on intraocular pressure. Invest.Ophth. 12, 204-212.

N. ANDERSON & A.M. ARTHURS (1976). Extremum principles and the elastic and fluid mechanical behavior of the human eye. Bull.Math.Biol. 38, 83-86.

M. ANLIKER (1972). Toward a nontraumatic study of the circulatory system. In Fung, Perrone & Anliker (1972), 337-379.

J.T. APTER (1966). Computer solution of non-linear differential equations of motion of ocular walls. 19th ACEMB, 195.

J.T. APTER (1967). Computer analysis of ocular tonograms. 7th ICMBE, Stockholm, 482.

M.F. ARMALY (1960). The effect of intraocular pressure on outflow facility. Arch. Ophth. 64, 125-132.

M.F. ARMALY (1964). On the consistency of tonography. Invest.Ophth. 3, 77-84.

M.F. ARMALY & A.H. HALASA (1963). The effect of external compression of the eye on intraocular pressure. I. Its variations with magnitude of compression and with age. Invest.Ophth. 2, 591-598.

M.F. ARMALY & N.C. JEPSON (1962). Accommodation and the dynamics of the steady-state intraocular pressure. Invest.Ophth. 1, 480-483.

J.D. ARONOWITZ & R.F. BRUBAKER (1976). Effect of intraocular gas on intraocular pressure. Arch.Ophth. 94, 1191-1196.

S.B. ARONSON, E.L. HOWES Jr., M.B. FISH, M. POLLYCOVE & D.N. O'DAY (1974). Ocular blood flow in experimentally induced immunogenic uveitis. Arch.Ophth. 91, 60-65.

P. BAILLIART (1917). La pression artérielle dans les branches de l'artère centrale de la rétine. Ann.d'Ocul. 154, 648-666.

E.H. BÁRÁNY (1947). The influence of intra-ocular pressure on the rate of drainage of aqueous humour: Stabilization of intra-ocular pressure or of aqueous flow? Brit.J.Ophth. 31, 160-176.

E.H. BÁRÁNY (1963). A mathematical formulation of intraocular pressure as dependent on secretion, ultrafiltration, bulk outflow, and osmotic reabsorption of fluid. Invest.Ophth. 2, 584-590.

E.H. BÁRÁNY (1967). Pseudofacility and uveo-scleral outflow routes. In Leydhecker (1967), 27-51.

B. BECKER & R.E. CHRISTENSEN (1956). Water-drinking and tonography in the diagnosis of glaucoma. Arch.Ophth. 56, 321-326.

B. BECKER & R.C. DREWS (1967). *Current concepts in ophthalmology*. C.V. Mosby, St. Louis.

B. BECKER & J.S. FRIEDENWALD (1953). Clinical aqueous outflow. Arch.Ophth. 50, 557-571.

C.H. BEDWELL (1966). The assessment of ocular tension and glaucoma. Ophth.Optician. 6, 914-921.

B. BENGTSSON (1972). Some factors affecting the distribution of intraocular pressures in a population. Acta.Ophth. 50, 33-46.

D.H. BERGEL (1961). The static elastic properties of the arterial wall. J.Physiol. 156, 445-457.

V. BERRY, S.M. DRANCE, R.L. WIGGINS & M. SCHULTZER (1966). A study of the errors of applanation tonometry and tonography on two groups of normal people. Can.J.Ophth. 1, 213-220.

M. BEST & M. BLUMENTHAL (1972). Elastic properties of intraocular blood vessels. Acta Ophth. 50, 458-468.

M. BEST, M. BLUMENTHAL, H.A. FUTTERMAN & M.A. GALIN (1969). Critical closure of intraocular blood vessels. Arch.Ophth. 82, 385-392.

M. BEST, M. BLUMENTHAL & M.A. GALIN (1970a). Distensibility of the intraocular vascular bed. Arch.Ophth. 84, 630-634.

M. BEST, D. GERSTEIN, N. WALD, A.Z. RABINOVITZ & G.H. HILLER (1973). Autoregulation of ocular blood flow. Arch.Ophth. 89, 143-148.

M. BEST, T.A. KELLY & M.A. GALIN (1970b). The ocular pulse - technical features. Acta Ophth. 48, 357-368.

M. BEST, S. MASKET & A.Z. RABINOVITZ (1971b). Measurement of vascular rigidity in the living eye. Arch.Ophth. 86, 699-705.

M. BEST, A.Z. RABINOVITZ, R. POLA & S. MASKET (1971a). Vascular distensibility in the eye. Arch.Ophth. 86, 88-93.

M. BEST & R. ROGERS (1974). Techniques of ocular pulse analysis in carotid stenosis. Arch.Ophth. 92, 54-58.

J.W. BETTMANN & V.G. FELLOWS Jr. (1956). A technique for the determination of blood-volume changes. Am.J.Ophth. 42, 161-167.

A. BILL (1962a). Intraocular pressure and blood flow through the uvea. Arch.Ophth. 67, 336-348.

A. BILL (1962b). Autonomic nervous control of uveal blood flow. Acta Physiol.Scand. 56, 70-81.

A. BILL (1963a). Blood pressure in the ciliary arteries of rabbits. Exp.Eye Res. 2, 20-24.

A. BILL (1963b). Effects of cervical sympathetic tone on blood pressure and uveal blood flow after carotid occlusion. Exp.Eye Res. 2, 203-309.

A. BILL (1967). Uveal circulation, a review of methods and results. In Leydhecker (1967). 52-72.

A. BILL (1969). Early effects of epinephrine on aqueous humor dynamics in vervet monkeys (Cerocopithecus ethiops). Exp.Eye Res. 8, 35-43.

A. BILL (1970). Ocular circulation. In Moses (1970a), 278-296.

A. BILL (1974). Effects of acetazolamide and carotid occlusion on the ocular blood flow in unanesthetized rabbits. Invest.Ophth. 13, 954-958.

A. BILL (1975). Blood circulation and fluid dynamics in the eye. Physiol.Rev. 55, 383-417.

J.N. BLOOM, R.Z. LEVENE, G. THOMAS & R. KIMURA (1976). Fluorophotometry and the rate of aqueous flow in man. I. Instrumentation and normal values. Arch.Ophth. 94. 435-443.

A. BORRÁS, A. MARTÍNEZ & M.S. MÉNDEZ (1969a). Ophthalmodynamometric and direct measurement of ophthalmic artery pressure. Am.J.Ophth. 67, 681-684.

A. BORRÁS, M.S. MÉNDEZ & A. MARTÍNEZ (1969b). Ophthalmic/brachial artery pressure ratio in man. Am.J.Ophth. 67, 684-688.

R.F. BRUBAKER (1976). Computer-assisted instruction of current concepts in aqueous humor dynamics. Am.J.Ophth. 82, 59-63.

R.F. BRUBAKER & D.M. WORTHEN (1973). The filtration coefficient of the intraocular vasculature as measured by low-pressure perfusion in a primate eye. Invest.Ophth. 12, 321-326.

C.J. BULPITT & C.T. DOLLERY (1971). Estimation of retinal blood flow by measurement of the mean circulation time. Cardiovasc.Res. 5, 406-412.

A.C. BURTON & S. YAMADA (1951). Relation between blood pressure and flow in the human forearm. J.Appl.Physiol. 4, 329-339.

H.G. BYNKE (1968). Influence of intraocular pressure on the amplitude of the corneal pulse. Acta Ophth. 46, 1135-1145.

H.G. BYNKE & B. SCHÉLE (1967). On the origin of the ocular pressure pulse. Ophthalmologica. 153, 29-36.

M.R. CHANDLER (1964). Aqueous flow measurements in man by the perilimbal suction cup technique. I. Observations in normal subjects and cases of glaucoma. Brit.J. Ophth. 48, 423-431.

P. CHAO & J.W. BETTMANN (1957). The relative volume of blood in the choroid and retina. Am.J.Ophth. 43, 294-295.

J.C.F. CHOW & J.F. APTER (1968). Wave propagation in a viscous imcompressible fluid contained in flexible viscoelastic tubes. J.Acous.Soc.Am. 44, 437-443.

D.F. COLE (1966). Aqueous humour formation. Doc.Ophth. 21, 116-238.

R. COLLINS (1978). Une nouvelle théorie globale de la lubrification des articulations. La Houille Blanche. 3/4, 211-219.

R. COLLINS & W.C.L. HU (1972). Dynamic deformation experiments on aortic tissue. J.Biomech. 5, 333-337.

H.P.G. DARCY (1856). *Les fontaines publiques de la ville de Dijon*. Victor Dalmont, Paris.

M. DAVANGER & Ø. HOLTER (1967). Intraocular pressure in non-equilibrium states. Acta Ophth. 45, 510-524.

M. DAVANGER & Ø. HOLTER (1971). Calculation of tonographic outflow facility. Acta Ophth. 49, 177-186.

H. DAVSON (1969). *The eye*. Vol. I. 2nd ed. Academic Press, New York.

S.M. DRANCE (1961). Relationship of consensual changes in intraocular pressure to arterial blood pressure. Arch.Ophth. <u>66</u>, 619-624.

R.C. DREWS (1967a). Tonography : introduction and history. In Becker & Drews (1967), 37-43.

R.C. DREWS (1967b). Clinical aspects of tonographic theory. In Becker & Drews (1967), 44-59.

W.S. DUKE-ELDER (1926). The ocular circulation : its normal pressure relationships and their physiological significance. Brit.J.Ophth. <u>10</u>, 513-572.

W.S. DUKE-ELDER & H. DAVSON (1948). The present position of the problem of the intra-ocular fluid and pressure. Brit.J.Ophth. <u>32</u>, 555-569.

W.S. DUKE-ELDER & J. GLOSTER (1968). *System of ophthalmology*. Vol. IV, Sect. A. *The physiology of the eye and of vision*. Henry Kimpton, London.

W.S. DUKE-ELDER & D.M. MAURICE (1957). Symbols of ocular dynamics. Brit.J.Ophth. <u>41</u>, 702-703.

W.S. DUKE-ELDER & K.C. WYBAR (1961). *System of ophthalmology*. Vol. II. *The anatomy of the visual system*. Henry Kimpton, London.

K.E. EAKINS (1969). A comparative study of intraocular pressure and gross outflow facility in the cat eye during anaesthesia. Exp.Eye Res. <u>8</u>, 106-115.

J.E. EISENLOHR & M.E. LANGHAM (1962). The relationship between pressure and volume changes in living and dead rabbit eyes. Invest.Ophth. <u>1</u>, 63-77.

J.E. EISENLOHR, M.E. LANGHAM & A.E. MAUMENEE (1962). Manometric studies of the pressure-volume relationship in living and enucleated eyes of individual human subjects. Brit.J.Ophth. <u>46</u>, 536-648.

R. ETIENNE (1965). *Les Glaucomes*. Diffusion Generale de Librairie, Marseille.

I. FATT (1975). Flow and diffusion in the vitreous body of the eye. Bull.Math.Biol. <u>37</u>, 85-90.

I. FATT & J. FORESTER (1972). Errors in eye tissue temperature measurements when using a metallic probe. Exp.Eye Res. <u>14</u>, 270-276.

D.H. FENDER & D.S. GILBERT (1966). Temporal and spatial filtering in the human visual system. Sci.Prog. <u>54</u>, 41-59.

H. FISCHER VON BÜNAU & F.P. FISCHER (1932). Hat der Glaskörper einen Stoffwechsel? Arch.Augenheilkd. <u>106</u>, 463-466.

M.B. FISH, S.B. ARONSON, M. POLLYCOVE & M.A. COON (1969). Ocular blood volume. Arch.Ophth. <u>82</u>, 377-380.

R.F. FISHER (1972). Value of tonometry and tonography in the diagnosis of glaucoma. Brit.J.Ophth. <u>56</u>, 200-204.

W. FLÜGGE (1962). *Handbook of engineering mechanics*. McGraw-Hill, New York.

M.C. FOX (1967). Continuous derivation of the pressure-flow relationship and outflow resistance for living human eyes from tonographic, manometric, and pressure cup pressure-decay curves. Exp.Eye Res. <u>6</u>, 243-260.

R.D. FREEMAN & I. FATT (1973). Environmental influences on ocular temperature. Invest.Ophth. <u>12</u>, 596-602.

J.S. FRIEDENWALD (1934). Retinal vascular dynamics. Am.J.Ophth. 17, 387-395.

J.S. FRIEDENWALD (1937). Contribution to the theory and practice of tonometry. Am.J.Ophth. 20, 985-1024.

J.S. FRIEDENWALD (1948). Some problems in the calibration of tonometers. Am.J.Ophth. 31, 935-944.

J.S. FRIEDENWALD (1949). The formation of the intraocular fluid. Am.J.Ophth. 32, 9-27.

J.S. FRIEDENWALD & B. BECKER (1955). Aqueous humor dynamics. Arch.Ophth. 54, 799-815.

A.B. FRIEDLAND (1978). A hydrodynamic model of aqueous flow in the posterior chamber of the eye. Bull.Math.Biol. 40, 223-235.

Y.C. FUNG (1967). Elasticity of soft tissues in simple elongation. Am.J.Physiol. 213, 1532-1544.

Y.C. FUNG, N. PERRONE & M. ANLIKER (1972). *Biomechanics: its foundations and objectives*. Prentice-Hall, Englewood Cliffs, New Jersey.

D. GAASTERLAND, C. KUPFER, K. ROSS & H.L. GABELNICK (1973). Studies of aqueous humor dynamics in man. III. Measurements in young normal subjects using norepinephrine and isoproterenol. Invest.Ophth. 12, 267-279.

M.A. GALIN, I. BARAS & R. CAVERO (1969a). Ophthalmodynamometry using suction. Arch. Ophth. 81, 494-500.

M.A. GALIN, I. BARAS, R. CAVERO & M. BEST (1969b). Compression and suction ophthalmodynamometry. Am.J.Ophth. 67, 388-392.

M.A. GALIN, I. BARAS & G.L. MANDELL (1961). Measurements of aqueous flow utilizing the perilimbal suction cup. Arch.Ophth. 66, 65-69.

M.A. GALIN, I. BARAS & J.M. McLEAN (1963). The technique of perilimbal suction cup analysis. Am.J.Ophth. 56, 883-888.

M.A. GALIN, M. BEST, G. PLECHATY & P. NUSSBAUM (1972). The ocular pulse. Trans.Am. Acad.Ophth.Otolaryng. 76, 1535-1541.

M.A. GALIN, M.L. KWITKO, J.M. DODICK & K.A. GITTER (1969c). Analysis of the ocular pulse. Can.J.Ophth. 4, 37-39.

D.H. GEROSKI & H.F. EDELHAUSER (1974). Metabolic evaluation of cryopreserved corneal tissue. Arch.Ophth. 91, 130-133.

J. GLOSTER (1966). *Tonometry and tonography*. J & A Churchill, London.

J. GLOSTER & E.S. PERKINS (1959). Distensibility of the human eye. Brit.J.Ophth. 43, 97-101.

J. GLOSTER & E.S. PERKINS (1963). The validity of the Imbert-Fick law as applied to applanation tonometry. Exp.Eye Res. 2, 274-283.

J. GLOSTER, E.S. PERKINS & M-L POMMIER (1957). Extensibility of strips of sclera and cornea. Brit.J.Ophth. 41, 103-110.

M.H. GOLDBAUM, M. SMITHLINE, T.A. POOLE & H.A. LINCOFF (1975). Geometric analysis of radial buckling. Am.J.Ophth. 79, 958-965.

H. GOLDMANN (1967). Summary. In Leydhecker (1967), 256-265.

H. GOLDMANN & T. SCHMIDT (1957a). Der Rigiditätskoeffizient (Friedenwald). Ophthalmologica. 133, 330-336.

H. GOLDMANN & T. SCHMIDT (1957b). Über Applanationstonometrie. Ophthalmologica. 134,
221-242.

H. GOLDMANN & T. SCHMIDT (1965). On applanation tonography. Ophthalmologica. 150,
65-75.

J.E. GOLDSTEIN, J.D. PECZON & D.G. COGAN (1965). Intraocular pressure and ophthalmodynamometry. Arch.Ophth. 74, 175-176.

W.P. GRAEBEL & G.W.H.M. van ALPHEN (1977). The elasticity of sclera and choroid of
the human eye, and its implication on scleral rigidity and accommodation. J.Biomech.
Eng. 99K, 203-208.

W.M. GRANT (1951). Clinical measurements of aqueous outflow. Arch.Ophth. 46, 113-131.

W.M. GRANT (1958). Further studies on facility of flow through the trabecular meshwork. Arch.Ophth. 60, 523-533.

K. GREEN & J.E. PEDERSON (1973). Aqueous humor formation. Exp.Eye Res. 16, 273-286.

H. HAGER (1964). Differential diagnosis of apoplexy by ophthalmodynamography.
Triangle, Sandoz J.Med.Sci. 6, 259-267.

R.W. HART (1970). Theory of nervous regulation of intraocular pressure. APL Technical Digest, Johns Hopkins Univ. 9, 2-13.

R.W. HART (1972). Theory of neural mediation of intraocular dynamics. Bull.Math.Bio-
phys. 34, 113-148.

R.W. HART & M.E. LANGHAM (1972). The diagnosis and adrenergic therapy of open-angle
glaucoma. Israel J.Med.Sci. 8, 1385-1393.

S.S. HAYREH (1963). Arteries of the orbit in the human being. Brit.J.Surg. 50,
938-953.

J. HETLAND-ERIKSEN (1966). On tonometry. 9. The pressure-volume relationship by
Schiøtz tonometry: concluding observations. Acta Ophth. 44, 893-900.

R.R. HIBBARD, C.S. LYON, M.D. SHEPHERD, E.H. McBAIN & W.K. McEWEN (1970). Immediate
rigidity of an eye. I. Whole, segments and strips. Exp.Eye Res. 9, 137-143.

J.B. HICKHAM & R. FRAYSER (1965). A photographic method for measuring the mean
retinal circulation time using fluorescein. Invest.Ophth. 4, 876-884.

T.H. HODGSON & R.K. MacDONALD (1957). Corneo-scleral tonography. Brit.J.Ophth. 41,
301-308.

M.G. HOLLAND, J. MADISON & W. BEAN (1960). The ocular rigidity function. Am.J.Ophth.
50, 958-974.

I. HØRVEN & H. NORNES (1971). Crest time evaluation of corneal indentation pulse.
Arch.Ophth. 86, 5-11.

B.N. HYMAN (1974). Doppler sonography: A bedside noninvasive method for assessment
of carotid artery disease. Am.J.Ophth. 77, 227-231.

M. ITOI, H. KANEKO & T. SUGIMACHI (1965). Anelasticity of eye ball and errors in
tonometry. Jap.J.Ophth. 9, 61-68.

H. KAUFMANN, H. FLOHR, W. BREULL, D. REDEL & H.W. DAHNERS (1973). Quantitative Messung der Durchblutungsgrösse des Auges. Arch.Klin.Exp.Ophth. 186, 181-190.

T. KENNER (1972). Flow and pressure in the arteries. In Fung, Perrone & Anliker (1972), 381-434.

V.E. KINSEY (1950). A unified concept of aqueous humor dynamics and the maintenance of intraocular pressure. Arch.Ophth. 44, 215-235.

V.E. KINSEY & E. BÁRÁNY (1949). The rate of flow of aqueous humor. II. Derivation of rate of flow and its physiologic significance. Am.J.Ophth. 32, 189-202.

A.S. KOBAYASHI, M.D. LARSON & A.F. EMERY (1977). Reevaluation of Woo's corneal and scleral data. *1977 Advances in Bioengineering*. Am.Soc.Mech.Eng., New York, 41-42.

A.S. KOBAYASHI, L.G. STABERG & W.A. SCHLEGEL (1973). Viscoelastic properties of human cornea. Exp.Mech. 13, 497-503.

A.S. KOBAYASHI, S.L-Y WOO, C. LAWRENCE & W.A. SCHLEGEL (1971). Analysis of the corneo-scleral shell by the method of direct stiffness. J.Biomech. 4, 323-330.

F.L.P. KOCH (1945). Ophthalmodynamometry. Arch.Ophth. 34, 234-247.

W. KORNBLUTH & E. LINNÉR (1955). Experimental tonography in rabbits. Arch.Ophth. 54, 717-724.

C.E.T. KRAKAU (1969). On the regulation of the intraocular pressure. Acta Ophth. 47, 1069-1088.

S. KRONHEIM, C.J. LAMBERTSEN, C. NICHOLS & P.L. HENDRICKS (1976). Inert gas exchange and bubble formation and resolution in the eye. In C.J. Lambertsen (Ed.), Proc. 5th Symp. Underwater Physiology, 327-334.

C. KUPFER & P. SANDERSON (1968). Determination of pseudo-facility in the eye of man. Arch.Ophth. 80, 194-196.

M.E. LANGHAM (1958). Aqueous humor and control of intra-ocular pressure. Physiol.Rev. 38, 215-242.

M.E. LANGHAM (1959a). Influence of the intra-ocular pressure on the formation of the aqueous humour and the outflow resistance in the living eye. Brit.J.Ophth. 43, 705-732. Abstract: J.Physiol. 143, 11P (1958).

M.E. LANGHAM (1959b). The effect of pressure on the rate of formation of the aqueous humour. J.Physiol. 147, 29P-30P.

M.E. LANGHAM (1962). Evaluation of the pressure cup technique for the measurement of aqueous humor formation. Invest.Ophth. 1, 484-492.

M.E. LANGHAM (1963). A new procedure for the analysis of intraocular dynamics in human subjects. Exp.Eye Res. 2, 314-324.

M.E. LANGHAM (1967). Manometric, pressure-cup, and tonographic procedures in the evaluation of intraocular dynamics. In Leydhecker (1967), 126-150.

M.E. LANGHAM & J.E. EISENLOHR (1963). A manometric study of the rate of fall of the intraocular pressure in the living and dead eyes of human subjects. Invest.Ophth. 2, 72-82.

M.E. LANGHAM, Y. KITAZAWA & R.W. HART (1971). Adrenergic responses in the human eye. J.Pharmacol.Exp.Ther. 179, 47-55.

M.E. LANGHAM & A.E. MAUMENEE (1964). The diagnosis and treatment of glaucoma based on a new procedure for the measurement of intraocular dynamics. Trans.Am.Acad. Ophth.Otolaryng. 68, 277-300.

R.J. LAST (1968). *Wolff's anatomy of the eye and orbit*. 6th ed. H.K. Lewis, London.

C. LAWRENCE & W.A. SCHLEGEL (1966). Ophthalmic pulse studies. I. Influence of intra-ocular pressure. Invest.Ophth. 5, 515-525.

H.A. LESTER (1966). Ocular oscillometry in cerebrovascular disease. Arch.Ophth. 76, 391-398.

R.Z. LEVENE (1957). Studies on ocular blood flow in the rabbit. Arch.Ophth. 58, 19-22.

R.Z. LEVENE (1961). Tonometry and tonography in a group health population. Arch. Ophth. 66, 42-47.

R.Z. LEVENE, J.N. BLOOM & R. KIMURA (1976). Fluorophotometry and the rate of aqueous flow in man. II. Primary open angle glaucoma. Arch.Ophth. 94, 444-447.

R. LEVENE & B. HYMAN (1969). The effect of intraocular pressure on the facility of outflow. Exp.Eye Res. 8, 116-121.

W. LEYDHECKER (1967). *Glaucoma*. Symp.Tutzing Castle 1966, S. Kargar, Basel/New York.

W. LEYDHECKER, K. AKIYAMA & H.G. NEUMANN (1958). Der intraokulare Druck gesunder menschlicher Augen. Klin.Monatsbl.Augenh. 133, 662-670.

E. LINNÉR (1959). The rate of aqueous flow in human eyes with and without senile cataract. Arch.Ophth. 61, 520-527.

A. LOBSTEIN & J. NORDMANN (1959). Modern ophthalmodynamometry. Doc.Ophth. 13, 397-430.

R.S. MACKAY, E. MARG & R. OECHSLI (1962). Arterial and tonometric pressure measure-ments in the eye. Nature. 194, 687-688.

F.J. MACRI (1961). Interdependence of venous and eye pressure. Arch.Ophth. 65, 442-449.

F.J. MACRI (1964). The intraocular and vascular pressure of the cat eye. Exp.Eye Res. 3, 266-282.

F.J. MACRI, T. WANKO & P.A. GRIMES (1958). The elastic properties of the human eye. Arch.Ophth. 60, 1021-1026.

F.J. MACRI, T. WANKO, P.A. GRIMES & L. von SALLMANN (1957). The elasticity of the eye. Arch.Ophth. 58, 513-519.

A.P. MAGITOT (1922). How to know the blood pressure in the vessels of the retina. Am.J.Ophth. 5, 777-784.

A.P. MAGITOT & P. BAILLIART (1922). Blood pressure in the vessels of the eye. Am.J. Ophth. 5, 824-828.

E. MAMMARELLA & M. MAIONE (1965). The pressure-volume relation in tonometry. Ophthalmologica. 149, 81-85.

S. MASKET, M. BEST, A.Z. RABINOVITZ & G. PLECHATY (1973). Vascular perfusion gradi-ents in the eye. Invest.Ophth. 12, 198-203.

E.H. McBAIN (1957). Tonometer calibration. Arch.Ophth. 57, 520-521.

E.H. McBAIN (1958). Tonometer calibration. II. Ocular rigidity. Arch.Ophth. 60, 1080-1091.

E.H. McBAIN (1960). Tonometer calibration. III. Volume of indentation and P_o determination. Arch.Ophth. 63, 936-942.

E.H. McBAIN, W.K. McEWEN & M. SHEPHERD (1967). An electrical model of the human eye. III. The model and the eye during suction cup procedure, and its reconciliation with tonography. Invest.Ophth. 6, 171-176.

D.A. McDONALD (1974). *Blood flow in arteries*. 2nd ed. Edward Arnold, London.

W.K. McEWEN (1958). Application of Poiseuille's law to aqueous outflow. Arch.Ophth. 60, 290-294.

W.K. McEwen (1967). Difficulties in measuring intraocular pressure and ocular rigidity. In Leydhecker (1967), 97-125.

W.K. McEWEN (1973). An advanced method of information processing for clinical Schiøtz tonograms. Invest.Ophth. 12, 834-838.

W.K. McEWEN & R. ST. HELEN (1965). Rheology of the human sclera. Unifying formulation of ocular rigidity. Ophthalmologica. 150, 321-346.

W.K. McEWEN, M. SHEPHERD & E.H. McBAIN (1967). An electrical model of the human eye. I. The basic model. Invest.Ophth. 6, 155-159.

M. MIKUNI, K. IWATA & H. TSUCHIYA (1961). Retinal blood flow. Jap.J.Ophth. 5, 272-277.

R.A. MOSES (1964). Effect of tonography on facility of outflow. Invest.Ophth. 3, 606-608.

R.A. MOSES (1970a). *Adler's physiology of the eye*. 5th ed. C.V. Mosby, St. Louis.

R.A. MOSES (1970b). Intraocular pressure. In Moses (1970a), 249-277.

R.A. MOSES (1970c). The aqueous. In Moses (1970a), 297-310.

R.A. MOSES (1971). The ciliary body in aphakia. Ann.Ophth. 3, 1112-1115.

R.A. MOSES (1972). A graphic analysis of aqueous humor dynamics. Am.J.Ophth. 73, 665-669.

C.C. MOW (1968). A theoretical model of the cornea for use in studies of tonometry. Bull.Math.Biophys. 30, 437-453.

R. NEUSCHÜLER & P. PEPE (1967). Premesse teorico - pratiche per lo studio di pazienti portatori di occlusioni carotidee mediante l'associazione dell'oftalmodinamometria e dell'oftalmodinamografia. Boll.Oculist. 46, 992-1001.

P. NIESEL (1969). Ophthalmodynamometrie. Ophthalmologica. 158, 342-352.

A.W. NYQUIST (1968). Rheology of the cornea: Experimental technique and results. Exp.Eye Res. 7, 183-188.

J.E. PEDERSON & K. GREEN (1973). Aqueous humor dynamics: A mathematical approach to measurement of facility, pseudofacility, capillary pressure, active secretion and x_c. Exp.Eye Res. 15, 265-276.

E.S. PERKINS & J. GLOSTER (1957a). Distensibility of the eye. Brit.J.Ophth. 41, 93-102.

E.S. PERKINS & J. GLOSTER (1957b). Further studies on the distensibility of the eye. Brit.J.Ophth. 41, 475-486.

R.B. PERRY & J.C. ROSE (1958). The clinical measurement of retinal arterial pressure. Circ. 18, 864-870.

A. PIRIE (1949). The effect of hyaluronidase injection on the vitreous humour of the rabbit. Brit.J.Ophth. 33, 678-684.

G.L. PORTNEY & F.J. SOUSA (1974). Negative findings on the seven-minute coefficient of outflow. Am.J.Ophth. 78, 848-851.

H.A. QUIGLEY & M.E. LANGHAM (1975). Comparative intraocular pressure measurements with the pneumatonograph and Goldmann tonometer. Am.J.Ophth. 80, 266-273.

R.D. RICHARDS & P.G. TITTEL (1973). Corneal and scleral distensibility ratio on enucleated human eyes. Invest.Ophth. 12, 145-151.

F. RIDLEY (1930). The intraocular pressure and drainage of the aqueous humour. Brit.J.Exp.Path. 11, 217-240.

C.E. RIVA, G.T. FEKE & I. BEN-SIRA (1978). Fluorescein dye-dilution technique and retinal circulation. Am.J.Physiol. 234, H315-H322.

J.W. ROHEN (1960). On the aqueous outflow resistance. Ophthalmologica. 139, 1-10.

G.L. RUSKELL (1961). Aqueous drainage paths in the rabbit. Arch.Ophth. 66, 861-870.

T. SAETEREN (1960a). Scleral rigidity in normal human eyes. Acta Ophth. 38, 303-311.

T. SAETEREN (1960b). Aqueous flow in hypotonic and glaucomatous eyes. Acta Ophth. 38, 347-363.

T. SAETEREN (1960c). The tonographic method for measuring aqueous flow. Acta Ophth. 38, 511-523.

R. ST. HELEN & W.K. McEWEN (1961). Rheology of the human sclera. I. Anelastic behaviour. Am.J.Ophth. 52, 539-548.

M.L. SALMON & A.J. GAY (1967). Ophthalmodynamography. Int.Ophth.Clin. 7, 745-764.

R.A. SCHIMEK (1964). *Simplified tonography.* V. Mueller, Chicago.

W.A. SCHLEGEL, C. LAWRENCE & L.G. STABERG (1972). Viscoelastic response in enucleated human eye. Invest.Ophth. 11, 593-599.

B. SCHWARTZ (1964). The effect of lid closure upon the ocular temperature gradient. Invest. Ophth. 3, 100-106.

B. SCHWARTZ (1965). Environmental temperature and the ocular temperature gradient. Arch.Ophth. 74, 237-243.

B. SCHWARTZ & M.R. FELLER (1962). Temperature gradients in the rabbit eye. Invest. Ophth. 1, 513-521.

N.J. SCHWARTZ, R.S. MACKAY & J.L. SACKMAN (1966). A theoretical and experimental study of the mechanical behavior of the cornea with application to the measurement of intraocular pressure. Bull.Math.Biophys. 28, 585-643.

M. SHEPHERD, E.H. McBAIN & W.K. McEWEN (1967). An electrical model of the human eye. II. The model and the eye during tonography. Invest.Ophth. 6, 160-170.

A. SILVERSTEIN, D. DONIGER & M.B. BENDER (1959). Manual compression of the carotid vessels, carotid sinus hypersensitivity and carotid artery occlusions. Ann.Int.Med 52, 172-181.

T.O. SIPPEL (1965). Energy metabolism in the lens during aging. Invest.Ophth. 4, 502-513.

H.A. STERN, K.G. WAKIM & C.W. RUCKER (1956). *In vivo* studies on the choroidal circu-lation of rabbits. Arch.Ophth. 56, 726-735.

E.B. STREIFF (1937). Le rapport entre la tension artérielle rétinienne et la pression générale. Schweiz.Med.Wschr. 67, 796.

G.C. STUCKEY (1974a). Tonometry and the elasticity of the eye. Austral.J.Ophth. 1, 112-113.

G.C. STUCKEY (1974b). Tonometry - the absolute determination of original intraocula pressure. Austral.J.Ophth. 2/3, 154-155.

E. SUVANTO, P. REISSELL & E. HIMANKA (1960). Retinal circulation time in man. Acta Ophth. 38, 46-49.

J. SZAPIRO & H. PAKULA (1963). Anatomical studies of the collateral blood supply to the brain and the retina. J.Neurol.Neurosurg.Psychiat. 26, 414-417.

J. SZAPIRO & I. ŚWIETLICZKO (1963). The significance of the retinal artery pressure in cerebrovascular insufficiency. J.Neurol.Neurosurg.Psychiat. 26, 410-413.

T. TANAKA, C. RIVA & I. BEN-SIRA (1974). Blood velocity measurements in human retin al vessels. Science. 186, 830-831.

W. THORBURN (1972a). Recordings of applanating force at constant intraocular pres-sure. I. Basic principle and apparatus description. Acta Ophth. 50, 720-736.

W. THORBURN (1972b). Recordings of applanating force at constant intraocular pres-sure. II. Procedures applied on living human eyes. Acta Ophth. 50, 887-895.

W. THORBURN (1973a). Recordings of applanating force at constant intraocular pres-sure. III. Intraocular volume-pressure relationship studied in intact human eyes. Acta Ophth. 51, 114-126.

W. THORBURN (1973b). Recordings of applanating force at constant intraocular pres-sure. IV. Intraocular volume changes due to changes in blood content. Acta Ophth. 51, 270-285.

W. THORBURN (1973c). Recordings of applanating force at constant intraocular pres-sure. V. Intraocular volume changes due to changes in content of aqueous humour. Acta Ophth. 51, 390-410.

S. TIMOSHENKO & J.N. GOODIER (1970). *Theory of elasticity.* 3rd ed. McGraw-Hill, New York.

R.H. TUFFIAS & M. ANLIKER (1967). Dynamic behavior of eye globes. SUDAAR No. 302, Stanford University Biomechanics Laboratory.

E.M. VAN BUSKIRK (1976). Changes in the facility of aqueous outflow induced by lens depression and intraocular pressure in excised human eyes. Am.J.Ophth. 82, 736-740.

E.M. VAN BUSKIRK & W.M. GRANT (1973). Lens depression and aqueous outflow in enucle ated primate eyes. Am.J.Ophth. 76, 632-639.

.M. VAN BUSKIRK & W.M. GRANT (1974). Influence of temperature and the question of involvement of cellular metabolism in aqueous outflow. Am.J.Ophth. 77, 565-572.

.J. VAN DER WERFF (1972a). Whole eye rigidity coefficients from segment experiments. Exp.Eye Res. 13, 181-183.

.J. VAN DER WERFF (1972b). The pressure measured in ophthalmodynamometry. Arch Ophth. 87, 290-292.

.A.J. VAN HEUVEN, A.B. MALIK, C.A. SCHAFFER, D. COHEN & M. MEHU (1977). Retinal blood flow derived from dye dilution curves. Arch.Ophth. 95, 297-301.

. VAUGHAN & T. ASBURY (1974). *General ophthalmology*. 8th ed. Lange Medical Publications. Los Altos, California.

.J. VIERNSTEIN & M. COWEN (1969). Static and dynamic measurements of the pressure-volume relationship in living and dead rabbit eyes. Exp.Eye Res. 8, 183-192.

.J. VIERNSTEIN & Y. KITAZAWA (1970). Measurements of factors affecting the precision of tonometry and tonography. Exp.Eye Res. 9, 91-97.

. WEIGELIN & A. LOBSTEIN (1963). *Ophthalmodynamometry*. Hafner Publishing, New York; Karger, Basel.

. WEINBAUM (1965). A mathematical model for the elastic and fluid mechanical behavior of the human eye. Bull.Math.Biophys. 27, 325-354.

. WEINBAUM (1977). A cellular conservation model for the replenishment of water and solute in epithelia with local osmotic transport. *1977 Bioengineering Symposium*. Am.Soc.Mech.Eng., New York, 47-49.

. WEINBAUM & J.R. GOLDGRABEN (1972). On the movement of water and solute in extra-cellular channels with filtration, osmosis and active transport. J.Fluid.Mech. 53, 481-512.

. WEINBAUM, M.E. LANGHAM, J.R. GOLDGRABEN & K. GREEN (1972). The role of secretion and pressure-dependent flow in aqueous humor formation. Exp.Eye Res. 13, 266-277

J.J. WEITER & J.T. ERNEST (1974). Anatomy of the choroidal vasculature. Am.J.Ophth. 78, 583-590.

J.J. WEITER, R.A. SCHACHAR & J.T. ERNEST (1973a). Control of intraocular blood flow. I. Intraocular pressure. Invest.Ophth. 12, 327-331.

J.J. WEITER, R.A. SCHACHAR & J.T. ERNEST (1973b). Control of intraocular blood flow. II. Effects of sympathetic tone. Invest.Ophth. 12, 332-334.

T.M. WILSON, R. STRANG, J. WALLACE, P.W. HORTON & N.F. JOHNSON (1973). The measurement of the choroidal blood flow in the rabbit using 85-Krypton. Exp.Eye Res. 16, 421-425.

V.G WONG & F.J. MACRI (1964). Vasculature of the cat eye. Arch.Ophth. 72, 351-358.

S.L-Y WOO, A.S. KOBAYASHI & C. LAWRENCE (1970). Nonlinear mechanical properties of intact human cornea and sclera. 23rd ACEMB, Washington, D.C.

S.L-Y WOO, A.S. KOBAYASHI, C. LAWRENCE & W.A. SCHLEGEL (1972b). Mathematical model of the corneo-scleral shell as applied to intraocular pressure-volume relations and applanation tonometry. Ann.Biomed.Eng. 1, 87-98.

S.L-Y WOO, A.S. KOBAYASHI, W.A. SCHLEGEL & C. LAWRENCE (1972a). Nonlinear material properties of intact cornea and sclera. Exp.Eye Res. 14, 29-39.

E.H. WOOD & R.P. HILL (1966). Thermography in the diagnosis of cerebrovascular occlusive disease. Acta Radiol. 5, 961-971.

D.F. WOODHOUSE (1969). A computer evaluation on tonography. Exp.Eye Res. 8, 127-142.

E. WUDKA & I.H. LEOPOLD (1956). Experimental studies of the choroidal vessels. Arch. Ophth. 55, 605-632; 857-885.

J. YTTEBORG (1960a). The role of intraocular blood volume in rigidity measurements on human eyes. Acta Ophth. 38, 410-436.

J. YTTEBORG (1960b). The effect of intraocular pressure on rigidity coefficient in the human eye. Acta Ophth. 38, 548-561.

J. YTTEBORG (1960c). Influence of bulbar compression on rigidity coefficient of human eyes *in vivo* and enucleated. Acta Ophth. 38, 562-577.

J. YTTEBORG (1960d). Further investigations of factors influencing size of rigidity coefficient. Acta Ophth. 38, 643-657.

SUBJECT INDEX

G appended to the page number denotes an entry in the Glossary.

amaurosis fugax 4

angiography 71, 73G

anterior chamber 2, 19, 42, 73G

applanation tonometer (Goldmann, MacKay-Marg) 56, 73G, 74G

aqueous humour 2, 43, 73G

    composition 3, 19

    dialysis 21, 74G

    diffusion 21, 24, 74G

    flow rate 19, 42, 43

    influence of anaesthetics and drugs 27, 43, 55

    outflow 20, 23, 41, 55, 58, 72

    production 20, 21, 41, 45, 55, 58, 72

    secretion 3, 14, 21, 22, 27, 76G

    ultrafiltration 21, 22, 26, 27, 76G

    uveoscleral bulk flow 23, 26, 27

    volume 3, 45

arterial distensibility 16

arterial pressure

    brachial 11, 65

    choroidal 11, 41, 73G

    circle of Hovius 11, 73G

    femoral 13

    ophthalmic 11, 41, 65, 69

    retinal 10, 11, 65

    uveal 10, 42

    wedge 12

autoregulation 14, 73G

axon 4, 28, 54, 73G

blackout 4

blindness iv, 20

blood

    flow 9, 42, 69

        dependence on arterial gases 9

    pressure regulation 10, 11

    volume 9, 14, 41, 42, 45

carotid compression 69

causal relationships 40, 72
ciliary body 3, 21, 23
circle of Hovius 11, 73G
circulation
      choroidal 8, 10, 14
      retinal 3, 8, 10, 14
      uveal 3, 8, 13, 14, 42
coefficient of ocular rigidity 29, 30, 42, 58, 73G
coefficient of vascular rigidity 18, 73G
collateralisation 69
constants, definition 7
cornea 2, 3, 73G
      curvature 3
      deformation 5, 35, 56, 59
      indentation 25, 59
corneo-scleral envelope 57, 73G
      thickness 2
      volume contained 29

Darcy's law 24
Doppler 10, 64, 71, 74G
dural sheath 5

elastic modulus 15, 75G
      pressure dependence 17
equilibrium state iv, 72
eye
      comparative dimensions 3
      diameter 2
      enucleated 30, 31, 35, 39, 74G
      volume 2, 29, 42, 45
facility 74G
      of aqueous production 22, 72
      of aqueous outflow 22, 25, 26, 28, 42, 43, 46, 50, 51, 58, 72
      pseudo- 25, 27, 55, 75G
      total 25, 28
      true 25, 27
filtration angle 23
filtration coefficient 26
fluorescein angiography, 12, 60, 73G
Fourier analysis 68

Friedenwald's coefficient 30, 72, 74G

glaucoma iv, 4, 20, 24, 28, 64, 74G
    incidence iv
Goldmann applanation tonometer 56, 74G

homeostasis 20, 42, 74G
hyaluronidase 24

indentation tonometer (Schiøtz) 56, 60, 61, 76G
index of refraction 3
influence diagram 40, 72
intraocular pressure v, 30, 41, 42, 45, 57, 65, 69, 72
    effects of gravity 6
    regulation 6, 19, 20, 29, 54
iris 2, 54, 74G

lamina cribrosa 4, 74G
lens 2, 19
logarithms 6
    base conversion 7

MacKay-Marg applanation tonometer 56
mucopolysaccharide 24

neural control 54, 72, 75G

occlusion 75G
    carotid artery 68
    retinal artery 66
ocular
    pulse analysis 68
    rigidity 29, 30, 41, 46, 50, 51, 72
        comparisons 36-38
ophthalmodynamometry 11, 12, 14, 65, 70, 75G

Poiseuille's law 25, 41, 65, 75G
Poisson's ratio 15, 75G
posterior chamber 2, 19, 42, 75G
pressure
    critical closing 12, 13, 16
    cutoff 21, 22, 45, 50, 51

episcleral venous 11, 12, 28, 41, 45, 46, 59, 74G
    perfusion 13
    pulsations 48
    transmural 12, 16, 41, 76G
    venous 11, 41
pseudofacility 25, 27, 55, 75G
pulse wave speed 18, 75G

radioactive tagging 9, 10
resistance to aqueous outflow 25, 26, 28, 54, 60
retina 2
    circulation time (RCT) 10, 42, 75G
    metabolic requirements 4
    nourishment 3

Schiøtz indentation tonometer 56, 60, 61, 76G
Schlemm's canal 23, 28, 76G
Standard Eye 42, 61, 76G
strain 15, 33, 76G
    volumetric 16, 34, 35
stress relaxation 36, 41, 45, 59, 72, 76G
stroke 64
suction cup procedure 41, 56, 62, 76G

tables
    eye dimensions 3
    aqueous characteristics 43
    vascular characteristics 44
thermography 70, 76G
time constants 35, 76G
tonogram 57, 76G
    processing 60
tonography 25, 27, 57, 69, 76G
tonometer 25, 41, 65
    applanation (Goldmann, MacKay-Marg) 56, 73G, 74G
    indentation (Schiøtz) 56, 60, 61, 76G
    suction cup 41, 56, 62, 76G
tonometry 29, 56, 57
trabecular network 23, 24, 27, 54, 76G

units 6

variables, definition 7
vascular 76G
     bed 2, 44
     dynamics 8
     rigidity 18, 46, 73G
vasoconstriction 9, 27, 55, 76G
vasodilation 9, 55, 77G
vasomotor tone 16, 54, 72, 77G
veins 77G
     collapse 9, 12
viscoelasticity 38, 45, 46, 62, 77G
vision 3, 4, 19
vitreous 2, 5, 29, 77G

wave speed 18, 75G

Bio-mathematics

Managing Editors: K. Krickeberg, S. A. Levin

Springer-Verlag
Berlin
Heidelberg
New York

Volume 8

A. T. Winfree

The Geometry of Biological Time

1979. Approx. 290 figures. Approx. 580 pages
ISBN 3-540-09373-7

The widespread appearance of periodic patterns
in nature reveals that many living organisms are
communities of biological clocks. This land-
mark text investigates, and explains in mathematical terms, periodic processes in living
systems and in their non-living analogues. Its
lively presentation (including many drawings),
timely perspective and unique bibliography will
make it rewarding reading for students and researchers in many disciplines.

Volume 9

W. J. Ewens

Mathematical Population Genetics

1979. 4 figures, 17 tables. XII, 325 pages
ISBN 3-540-09577-2

This graduate level monograph considers the
mathematical theory of population genetics,
emphasizing aspects relevant to evolutionary
studies. It contains a definitive and comprehensive discussion of relevant areas with references
to the essential literature. The sound presentation and excellent exposition make this book a
standard for population geneticists interested in
the mathematical foundations of their subject
as well as for mathematicians involved with
genetic evolutionary processes.

Volume 10

A. Okubo

Diffusion and Ecological Problems: Mathematical Models

1980. 114 figures. XIII, 254 pages
ISBN 3-540-09620-5

This is the first comprehensive book on mathematical models of diffusion in an ecological
context. Directed towards applied mathematicians, physicists and biologists, it gives a
sound, biologically oriented treatment of the
mathematics and physics of diffusion.

Journal of

Mathematical Biology

ISSN 0303-6812　　　　　　　　　　　　Title No. 285

Editorial Board:
H. T. Banks, Providence, RI; **H. J. Bremermann,**
Berkeley, CA; **J. D. Cowan,** Chicago, IL; **J. Gani,**
Canberra City; **K. P. Hadeler** (Managing Editor),
Tübingen; **S. A. Levin** (Managing Editor), Ithaca, NY;
D. Ludwig, Vancouver; **L. A. Segel,** Rehovot; **D. Varjú,**
Tübingen

Advisory Board: M. A. Arbib, W. Bühler, B. D. Coleman,
K. Dietz, F. A. Dodge, P. C. Fife, W. Fleming, D. Glaser,
N. S. Goel, S. P. Hastings, W. Jäger, K. Jänich, S. Karlin,
S. Kauffman, D. G. Kendall, N. Keyfitz, B. Khodorov,
J. F. C. Kingman, E. R. Lewis, H. Mel, H. Mohr,
E. W. Montroll, J. D. Murray, T. Nagylaki, G. M. Odell,
G. Oster, L. A. Peletier, A. S. Perelson, T. Poggio,
K. H. Pribram, J. M. Rinzel, S. I. Rubinow, W. v. Seelen,
W. Seyffert, R. B. Stein, R. Thom, J. J. Tyson

The **Journal of Mathematical Biology** publishes papers
in which mathematics leads to a better understanding
of biological phenomena, mathematical papers inspired
by biological research and papers which yield new experimental data bearing on mathematical models. The
scope is broad, both mathematically and biologically
and extends to relevant interfaces with medicine,
chemistry, physics and sociology. The editors aim to
reach an audience of both mathematicians and
biologists.

Springer-Verlag
Berlin
Heidelberg
New York

Subscription information and sample copy
upon request.